Interpreting NMR Spectra Is *Not* Impossible!

Nuclear magnetic resonance (NMR) spectroscopy is the gold standard for the identification of reaction products and unknown substances in the synthetic laboratory. This book is designed for Organic Chemistry lecture and/or laboratory courses at the high school or college and university level. There is a deliberate and gradual scaffolding of complexity that divides the multitude of information contained in an NMR spectrum into easily understood pieces that are introduced one at a time with ample opportunities for practice at each stage. The topics include an introduction to the theory of NMR, several chapters on proton NMR (the number of signals, chemical shift, integration, splitting, and advanced topics), and carbon-13 NMR. The book culminates with twenty additional, comprehensive problems. All of the problems throughout the text are accompanied by fully worked solutions with detailed explanations. The mission of this volume parallels that of the previous book in the collection: *Calculating Theoretical Yield Is* Not *Impossible!* Future volumes will address additional Organic Chemistry topics.

Michael S. Leonard earned his B.A. in Chemistry from Goucher College in 1998 under the direction of Professor David E. Horn. He then attended the University of Pennsylvania for his doctoral studies in the laboratory of Professor Madeleine M. Joullié. After earning his Ph.D. in 2003, he joined the faculty of Washington & Jefferson College, where he is a chemistry professor.
https://orgowithleonard.com

Interpreting NMR Spectra Is *Not* Impossible!

A Primer for an Organic Chemistry Course

Michael S. Leonard

CRC Press
Taylor & Francis Group
Boca Raton London New York

CRC Press is an imprint of the
Taylor & Francis Group, an **informa** business

SUPPORT MATERIAL

Designed cover image: Shutterstock

First edition published 2027
by CRC Press
2385 NW Executive Center Drive, Suite 320, Boca Raton FL 33431

and by CRC Press
4 Park Square, Milton Park, Abingdon, Oxon, OX14 4RN

CRC Press is an imprint of Taylor & Francis Group, LLC

ISBN: 978-1-041-32526-0 (hbk)
ISBN: 978-1-041-32525-3 (pbk)
ISBN: 978-1-003-78332-9 (ebk)

DOI: 10.1201/9781003783329

Typeset in Times
by SPi Technologies India Pvt Ltd (Straive)

Access the Instructor and Student Resources: \\EUW1FS01\bks_digipub\COMPANION_WEBSITE_HANDOVERS\EDITORIAL\9781041325260_Leonard

Dedication

To all those who toil in relative obscurity to educate the next generation so that they can make their way in the world.

Contents

About the Author

Michael S. Leonard earned his B.A. in Chemistry from Goucher College in 1998 under the direction of Professor David E. Horn. He then attended the University of Pennsylvania for his doctoral studies in the laboratory of Professor Madeleine M. Joullié. After earning his Ph.D. in 2003, he joined the faculty of Washington & Jefferson College, where he is a chemistry professor.

Preface

The first book in this series—Calculating Theoretical Yield Is *Not* Impossible!—is a primer that guides students through a fundamental synthetic laboratory skill: the calculation of theoretical and percent yield. Complexity is introduced gradually through short chapters and a manageable number of practice problems, with fully explained answers, at each stage. This second book in the series applies the same methodology to the gold standard for identification of reaction products and unknown substances in the synthetic laboratory: nuclear magnetic resonance (NMR) spectroscopy.

Organic Chemistry occupies a spot in the public consciousness as a terribly hard subject, and NMR spectroscopy is arguably the apex of its difficulty. This is so much the case that the movie *Good Will Hunting* showcased the eponymous protagonist's prodigious intellectual talents by having him cavalierly assign the proton spectrum for ibogamine. In reality though, the interpretation of NMR spectra does not require Will Hunting's legendary capabilities. All it necessitates is a level of understanding that is born of clear explanation with a deliberate scaffolding of topics and a methodical approach to problem solving. This book provides the tools for all students willing to devote the effort to master this traditionally intimidating topic.

Information is introduced one layer at a time using short chapters, each with a manageable number of practice problems. We begin with an introduction to the theory of NMR and a focus on proton NMR. Then, there is a discussion of the number of signals that a compound will produce. This is followed by an analysis of chemical shift, but the examples and problems utilize spectra that only exhibit singlets (i.e., signals consisting of one spike) so that there is no confusing splitting (i.e., multi-spiked signals) before that topic has been introduced. The chapter on integration introduces peak shape as a complicating factor when determining how to compare the size of signals, and the subsequent chapter explains multiplicity in detail. The graphics are designed for maximum clarity, for instance emphasizing how a methine (CH), methylene (CH_2), or methyl (CH_3) group with the same number of neighbors will yield the same splitting pattern and will differ only in their integration. Now that it is appropriate, spectra with splitting are utilized in examples and problems. This is followed by chapters on advanced topics (such as complex splitting patterns and diastereotopic methylene hydrogens) and on carbon-13 NMR. Each chapter contains practice problems with detailed solutions, and the book concludes with a chapter of twenty culminating practice problems and a separate chapter with fully explained solutions to each of these questions.

While you may have heard other students bemoan the fact that NMR is impossible, I assure you that it is not. In fact, interpreting NMR spectra can be a relatively straightforward process once you truly understand the basic tenets and have practiced them in a variety of scenarios. This primer will help you do that. You've got this! Let's get to work.

1 NMR
An Introduction

Nuclear magnetic resonance (NMR) spectroscopy is the gold standard for structural determination in organic chemistry. Any form of spectroscopy uses light to study matter. NMR is unique in that it exploits nuclear spin to provide information about certain nuclei in a molecule using radio-frequency light. Nuclei with an odd atomic number and/or an odd mass number have a nuclear spin and are therefore NMR active, meaning that we can learn something about these nuclei using the technique. The most commonly studied nuclei are ^{1}H and ^{13}C, though there are others that can be probed in this way (e.g., ^{14}N, ^{19}F). In this chapter, we'll discuss how the technique works in general terms, and the rest of the book will be devoted to what is most important in organic chemistry: interpreting NMR spectra.

As noted above, several types of nuclei are NMR active. The most common type of NMR spectroscopy is proton NMR. The term "proton" in this context refers to the ^{1}H nucleus, which literally possesses only a single proton. This technique may also be called ^{1}H NMR. The method formally investigates the hydrogen nucleus (a proton), so the term "proton" may be used to refer to a particular hydrogen atom. In other words, the terms "proton" and "hydrogen" are sometimes used interchangeably when discussing this type of NMR, even though there is a difference of one electron between the hydrogen nucleus and the hydrogen atom.

Since proton NMR is the most common type of NMR, we'll refer to that particular nucleus in the discussion that follows. A hydrogen nucleus (or proton) has a spin that is randomly oriented under normal circumstances (Figure 1.1).

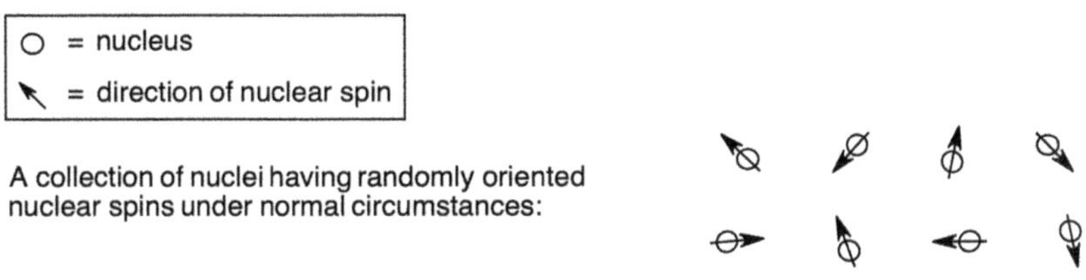

FIGURE 1.1 Nuclear spins are randomly oriented under normal circumstances.

However, when a sample is placed in a magnetic field, the nuclear spin is no longer randomly oriented. It can now have only two spin states: aligned with the external field or aligned against the external field. Aligning with the field is lower in energy, so the vast majority of nuclei will be in this so-called ground state (Figure 1.2).

DOI: 10.1201/9781003783329-1

FIGURE 1.2 In an external magnetic field, nuclear spins can only align with or against the external field.

Radio-frequency light happens to be the segment of the electromagnetic spectrum with the right amount of energy to flip the nuclear spin to the higher-energy (or excited) state in which the nuclear spin opposes the external field (Figure 1.3).

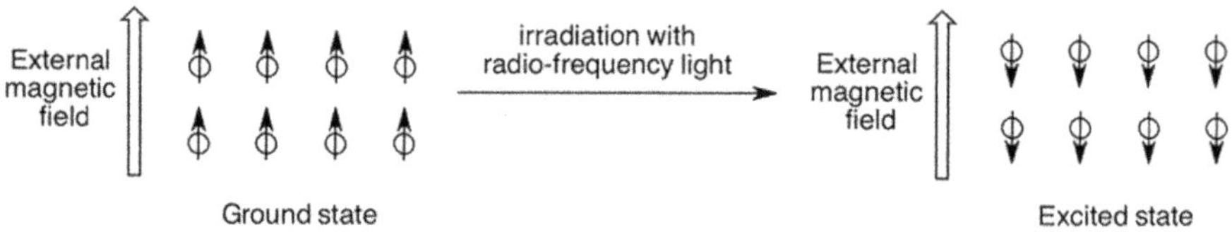

FIGURE 1.3 When irradiated with radio-frequency light, the nuclear spins are flipped to the excited state.

A strong external magnetic field will create a larger energy difference between spin up and spin down, but a weak external field will create only a small energy difference between these two states (Figure 1.4).

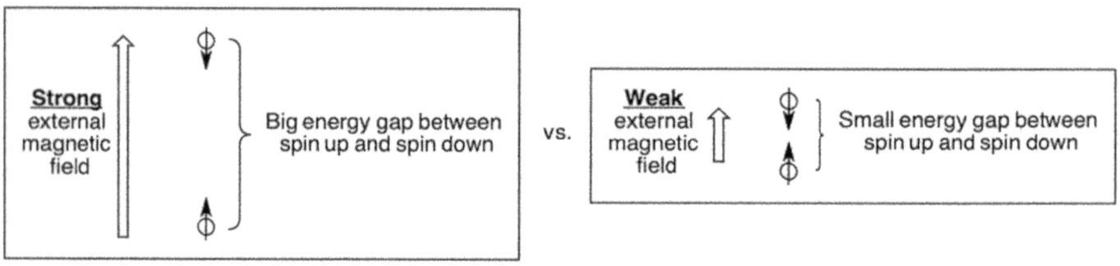

FIGURE 1.4 The difference in energy between spin up and spin down depends on the strength of the external field.

When irradiation with radio-frequency light stops, the nuclei relax back to the ground state, but because energy is always conserved, the difference in energy between the excited and ground states must go somewhere. It is emitted as radio-frequency light (Figure 1.5).

Therefore, when treated this way, protons in a molecule will provide evidence of their existence. However, if this were the full extent of the story, it wouldn't be very interesting since we know that the overwhelming majority of organic molecules contain protons. Simply knowing that protons are present is not especially informative. What makes proton NMR extraordinarily powerful in structural

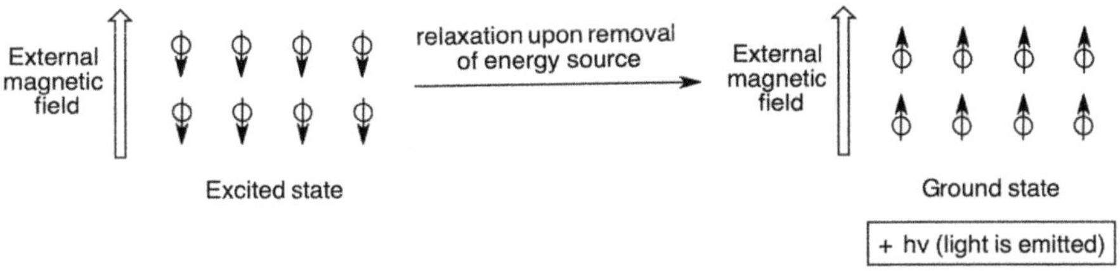

FIGURE 1.5 When irradiation stops, the nuclei relax back to the ground state and emit radio-frequency light.

determination is that protons in different chemical environments will yield different signals. Therefore, we won't simply receive information telling us that protons are present. We will receive information about each unique type of proton in the molecule.

This is possible because electrons induce a magnetic field that opposes the external field, thereby "shielding" the nucleus from the full brunt of the external magnetic field. Let's compare two very different types of protons. One has a great deal of electron density around it, so the electrons induce a powerful field opposing the external one. Therefore, the effective field "felt" by the nucleus is much smaller, and the spin-up to spin-down transition is low in energy. This proton was heavily shielded from the external field (Figure 1.6).

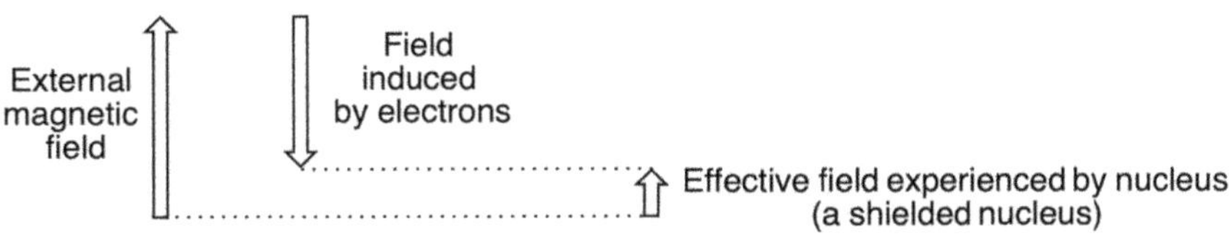

FIGURE 1.6 A proton surrounded by a great deal of electron density is shielded.

Now, consider a second type of proton that has little electron density around it. With little electron density, the induced field will be small, and the effective field experienced by the nucleus will be almost as great as the external field (Figure 1.7). This nucleus was not well shielded. For the sake of contrast, this situation is sometimes referred to as deshielded.

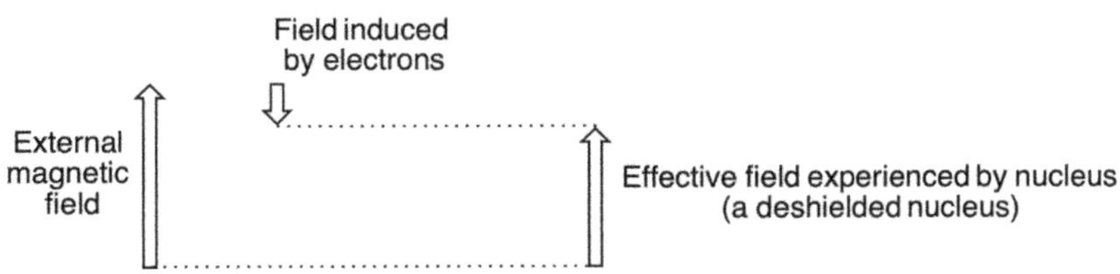

FIGURE 1.7 A proton surrounded by little electron density is deshielded.

In an organic molecule, there are often many different chemical environments in which protons can find themselves. A simplistic example would be that protons at different distances from a functional group will be in different chemical environments. Each of these types of protons has a unique amount of electron density around it. It therefore experiences a unique effective field and creates its own signal in the NMR spectrum. As a result, we'll obtain a great deal of information about the molecule's structure from this technique.

2 Proton NMR Part I
The Number of Signals

The first and most basic piece of information in a nuclear magnetic resonance (NMR) spectrum is the number of signals. In proton NMR, each proton in a chemically unique environment yields a unique signal. However, protons that are in identical chemical environments will be lumped together into a single signal. So, we need a strategy to determine if protons are in the same or different chemical environments.

Here's a simple exercise you can use to make the determination. Imagine that you are on the phone with a friend who takes organic chemistry at another school. You both know the exact same amount of organic chemistry. Can you describe the difference between two protons to your friend without using the words "up," "down," "left," or "right?" If so, the protons are in chemically unique environments. If not, they are in identical chemical environments. You cannot use the words "up," "down," "left," or "right" when attempting to describe a difference because molecules are always tumbling around in solution. Therefore, "up" can become "down" and "left" can become "right" in the blink of an eye.

Let's try this exercise with the protons found in methane (CH_4, Figure 2.1).

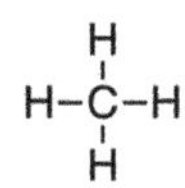

FIGURE 2.1 Methane.

Since all four hydrogen atoms are bonded to the same carbon atom, there is no difference between them, except for where they are pointing in space (i.e., up, down, left, and right). If the molecule is flipped over, the hydrogen on the top would trade places with the hydrogen on the bottom, or the hydrogen on the left would trade places with the hydrogen on the right. Therefore, these differences are not meaningful. The four hydrogens are in identical chemical environments and can be considered to be of the same type. With only one type of hydrogen atom in the molecule, we expect only one signal in the proton NMR spectrum of methane.

Let's add one carbon atom to the backbone to produce ethane (Figure 2.2) and repeat the exercise.

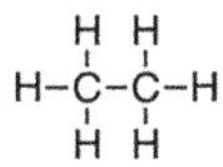

FIGURE 2.2 Ethane.

DOI: 10.1201/9781003783329-2

Once again, the hydrogen atoms are indistinguishable. Any of the six hydrogens you might choose can be described as a hydrogen that is part of a methyl (CH_3) group that is adjacent to another methyl group. Although there are six hydrogen atoms in the molecule, there is only one type of hydrogen atom, and we can expect only one signal in the proton NMR spectrum.

When we add one more carbon atom to the backbone to obtain propane (Figure 2.3), the situation becomes more interesting.

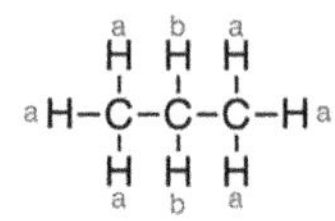

FIGURE 2.3 Propane.

Now, there are two types of hydrogen. The six hydrogens of the methyl groups are labeled as type "a." They are all indistinguishable from one another due to the molecule's symmetry. However, the two hydrogens of the methylene (CH_2) group comprise a different type of proton and are consequently labeled "b." Being part of a methyl group is different than being part of a methylene group, so these types of hydrogens can be differentiated from one another. We expect the proton NMR spectrum of propane to contain two signals.

Symmetry plays a key role in the outcome of the previous scenario. Let's now turn our attention to a less symmetrical molecule (Figure 2.4).

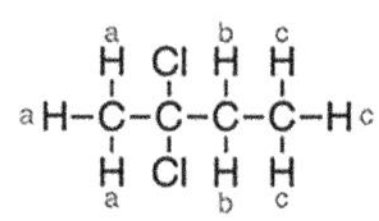

FIGURE 2.4 2,2-Dichlorobutane.

This molecule also contains two methyl groups and one methylene. However, it will yield a different number of signals in the proton NMR spectrum. The hydrogens of the methylene group (type "b") are clearly different from the hydrogens of the methyl groups. This time, though, the two methyl groups are not identical to one another. The hydrogens of methyl group "a" are different than the hydrogens of methyl group "c" because methyl group "a" is closer to the chlorine atoms. Any methyl-group hydrogen of type "a" is three bonds from a chlorine atom, while any methyl-group hydrogen of type "c" is four bonds from a chlorine atom. Consequently, this molecule has three types of hydrogens and will give a proton NMR spectrum with three signals.

Since molecules are frequently drawn using skeletal structures, let's employ that method to illustrate a slightly larger compound (Figure 2.5).

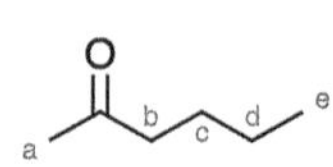

FIGURE 2.5 2-Hexanone.

This ketone, 2-hexanone, has two methyl groups and three methylene groups, but each group is unique. When comparing the hydrogen atoms of the two methyl groups, type "a" is much closer to the carbonyl (C=O) than type "e." When comparing the methylene groups (b–d), each is at a different distance from the carbonyl. Therefore, this molecule contains five types of hydrogen atoms and will produce five signals in the proton NMR spectrum.

In our next case study, let's consider the aromatic molecule toluene (Figure 2.6).

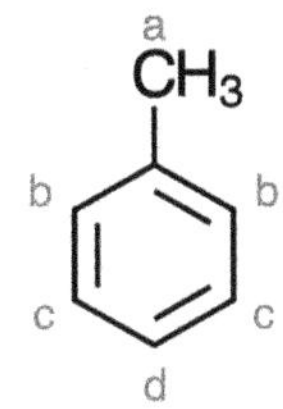

FIGURE 2.6 Toluene.

In this case, you might be tempted to say that there are two types of hydrogens: those of the methyl group and those on the aromatic ring. However, hydrogen atoms on a ring can be in different chemical environments. In this case, the ring hydrogens labeled "b" are *ortho* (adjacent) to the methyl substituent. The hydrogens labeled "c" are *meta* (one carbon further removed), and the hydrogen labeled "d" is *para* (opposite) to the substituted position. There are five hydrogen atoms on the aromatic ring. Due to the axis of symmetry that slices through the ring vertically, these five hydrogens are sorted into three different types (b–d). Naturally, the hydrogens of the methyl group (a) are different than those on the ring, so we have a total of four types of hydrogens in toluene. Its proton NMR spectrum will show four signals.

Finally, let's consider the role that restricted rotation plays in determining the types of hydrogens in a molecule (Figure 2.7).

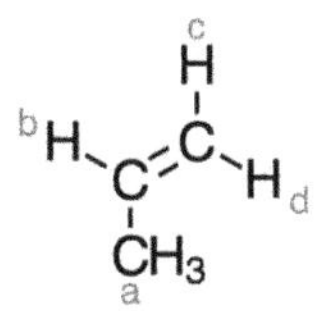

FIGURE 2.7 Propylene.

Propylene contains a carbon-carbon double bond, and alkenes (C=C) cannot freely rotate about their double bond. As a result, despite hydrogens "c" and "d" being bonded to the same carbon atom, they are in different chemical environments. Hydrogen "d" is always held closer (*cis*) to the methyl group (a), while hydrogen "c" is always further from (*trans* to) the methyl group. This places them in distinct chemical environments and causes them to produce different NMR signals. The methine (CH) hydrogen (b) and the methyl hydrogens (a) are also separate types yielding their own signals, so this molecule will produce four signals in the proton NMR.

CHAPTER 2 PRACTICE PROBLEMS

1. For each of the following structures (Figures 2.8–2.11), label the different types of hydrogen atoms with letters and predict the total number of signals in the proton NMR.
   (a)

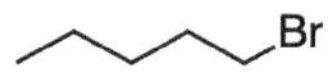

FIGURE 2.8 1-Bromopentane.

(b)

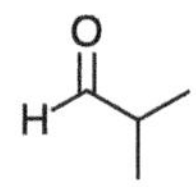

FIGURE 2.9 Isobutyraldehyde.

(c)

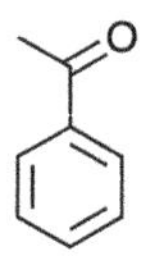

FIGURE 2.10 Acetophenone.

(d)

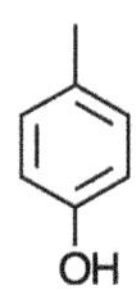

FIGURE 2.11 *p*-Cresol.

2. For each of the following, draw a structure consistent with the information provided.
   (a) An alkyl fluoride having the formula C_3H_7F that would give three proton NMR signals.
   (b) A ketone having the formula $C_6H_{12}O$ that would give only two proton NMR signals.
   (c) An aromatic compound with the formula C_7H_7Cl that shows three signals in the proton NMR.
   (d) An alkene with the formula C_5H_{10} that produces four signals in the proton NMR spectrum.

SOLUTIONS TO CHAPTER 2 PRACTICE PROBLEMS

1.

(a) 1-Bromopentane (Figure 2.12) contains one methyl group and four methylene groups. The methyl group (a) is clearly unique. Less obviously, the four methylenes (b–e) are all distinct from one another because they are all at different distances from the halogen. Therefore, 1-bromopentane will cause five signals in the proton NMR.

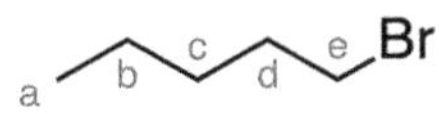

FIGURE 2.12 1-Bromopentane with the different types of hydrogens labeled.

(b) The hydrogen of the aldehyde (a) in isobutyraldehyde (Figure 2.13) stands out as clearly different from all the other hydrogens in the molecule. The methine labeled "b" is unique as well; however, the two methyl groups (c) both stem from the same carbon and are therefore identical to one another. Consequently, isobutyraldehyde will give three signals in the proton NMR.

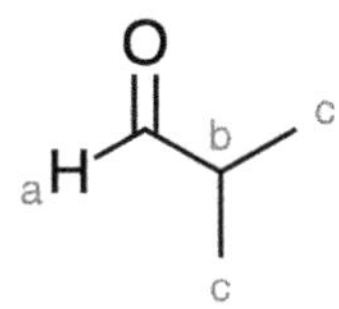

FIGURE 2.13 Isobutyraldehyde with the different types of hydrogens labeled.

(c) Acetophenone (Figure 2.14) contains only one methyl group (a). Those hydrogens constitute one type. The aromatic ring bears five hydrogen atoms but of only three different types: those *ortho* to the substituent (b), those *meta* to the substituent (c), and the one *para* to the substituent (d). Acetophenone produces a proton NMR spectrum with four signals.

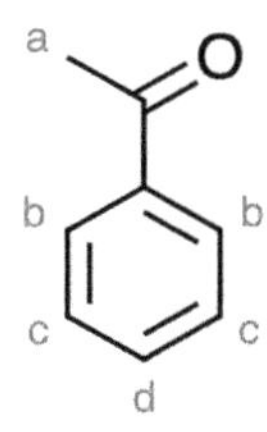

FIGURE 2.14 Acetophenone with the different types of hydrogens labeled.

(d) *p*-Cresol (Figure 2.15) contains a single methyl group (a). The hydrogen of the hydroxyl group (d) also clearly stands out as unique.

The aromatic ring has an axis of symmetry cutting through the center of the ring vertically. As such, the two hydrogens labeled "b" are identical to one another; they are both *ortho* to the methyl group and *meta* to the hydroxyl group. Similarly, the two hydrogens labeled "c" are identical to one another; they are both *ortho* to the hydroxyl group and *meta* to the methyl group. *p*-Cresol yields four signals in its proton NMR.

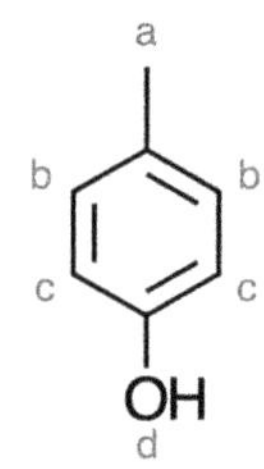

FIGURE 2.15 *p*-Cresol with the different types of hydrogens labeled.

2.

(a) There are two constitutional isomers for this molecular formula: 1-fluoropropane and 2-fluoropropane. However, in order to produce three signals in the proton NMR, there can be no symmetry in the molecule, so 1-fluoropropane (Figure 2.16) is the correct choice.

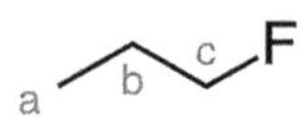

FIGURE 2.16 1-Fluoropropane.

(b) There are multiple constitutional isomers for a ketone having the molecular formula $C_6H_{12}O$. You do *not* want to start randomly drawing isomers, hoping to stumble upon a correct choice. It's much better to think strategically. There are a fair number of hydrogens (twelve), but only two signals to account for all of them. This indicates segments of the molecule having a high level of symmetry. For instance, a *tert*-butyl group has three methyls (b) stemming from the same carbon. Therefore, all nine of its hydrogen atoms will be identical and produce only one signal. This would leave only three hydrogens, enough for one more methyl group (a) that is different from the others because it originates from a different location. Since a ketone requires two R groups on the carbonyl carbon, placing a methyl and a *tert*-butyl group on the carbonyl carbon provides an isomer fitting the criteria outlined in the question (Figure 2.17).

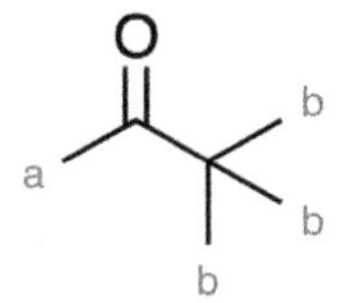

FIGURE 2.17 3,3-Dimethyl-2-butanone.

(c) We are told that the answer should be an aromatic compound, so we know that we're working with a substituted benzene ring. A benzene ring accounts for six carbons, leaving one carbon and the chlorine that need to be placed as substituents on this ring. Once again, we want to consider symmetry as a useful tool to account for seven hydrogen atoms with only three signals. If the carbon and chlorine are placed opposite (*para* to) one another on the ring (Figure 2.18), then the compound has one unique methyl group (a), a pair of aromatic hydrogens (b) that are *ortho* to the methyl and *meta* to the chlorine, and a pair of aromatic hydrogens (c) that are *ortho* to the chlorine and *meta* to the methyl group.

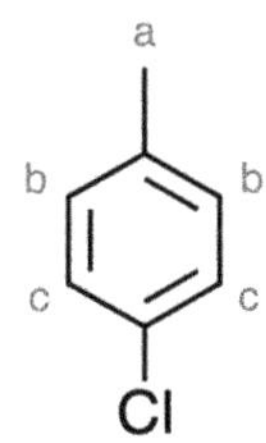

FIGURE 2.18 *p*-Chlorotoluene.

(d) In this question, we are dealing with a five-carbon alkene. However, the three isomers of a linear five-carbon alkene (1-pentene, *cis*-2-pentene, and *trans*-2-pentene) all produce too many signals. Consequently, we are forced to turn our attention to a substituted butene. 2-Methyl-2-butene (Figure 2.19) fits the bill. It contains three methyl groups (a, c, and d), but they are all different from one another. Methyl group "a" is unique because it is adjacent to a methine. While methyl groups "c" and "d" both emanate from the same carbon, the restricted rotation of the carbon-carbon double bond renders them distinct. Methyl group "d" is *cis* to methyl "a," while methyl group "c" is *trans* to "a." The methine (b) is the fourth signal for this molecule.

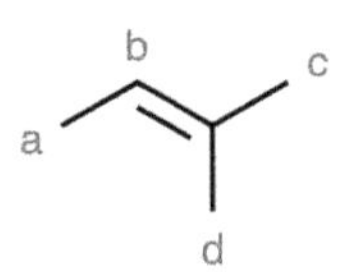

FIGURE 2.19 2-Methyl-2-butene.

3 Proton NMR Part II
Chemical Shift

The second significant piece of information provided by nuclear magnetic resonance (NMR) is called the chemical shift. Essentially, this refers to the location of a signal on the x-axis, but we need a little bit of background to understand why the location on the x-axis is termed chemical shift.

The difference in energy between spin states depends on the strength of the external magnetic field, and different NMR spectrometers have different magnetic field strengths. Therefore, if we reported spectra on an x-axis that simply referred to the energy difference in some way—for instance, using energy itself or using frequency, which is directly proportional to energy—then the location of signals would be dependent upon the specific instrument used to acquire the spectrum. That would not be acceptable because spectra acquired on different instruments could not be easily compared. Consequently, we need to normalize for (or account for) the role of magnetic field strength.

This is accomplished using an internal standard. NMR samples typically have a very tiny amount of a known substance added to them. This known substance is called an internal standard, and it is effectively a point of reference. The internal standard most commonly used is tetramethylsilane, which has the formula $(CH_3)_4Si$ and is often abbreviated as TMS. All of its hydrogens are chemically equivalent, so it produces only one signal. By convention, we agree that the TMS signal marks the zero point on the x-axis. To find the chemical shift or x-axis value for other signals, the following formula is used:

$$\delta(\text{ppm}) = \frac{\nu \text{ for signal}(\text{Hz}) - \nu \text{ for TMS}(\text{Hz})}{\text{operating frequency of the magnet}(\text{MHz})} \times 10^6$$

Chemical shift is denoted by the lowercase Greek letter delta (δ). It is measured in parts per million (ppm) for reasons that will become clear momentarily. The numerator is the difference between the frequency (ν) of a signal, measured in Hertz (Hz), and the frequency of the TMS signal, also measured in Hz. This difference is divided by the operating frequency of the magnet in megahertz (MHz), which is merely one million Hz. The quotient is multiplied by 10^6 to cancel out that factor of one million, therefore giving a chemical shift in parts per million.

You don't actually need to do these calculations. The computer software does them for you. The point of mentioning the formula is merely to explain why we have an x-axis labeled chemical shift and to justify the presence of TMS as an internal standard in our NMR samples.

DOI: 10.1201/9781003783329-3

By acquiring a multitude of proton NMR spectra over the years, the chemical community has learned that protons in particular chemical environments show up in predictable places on the x-axis. Figure 3.1 shows the correlation between chemical environment and chemical shift. Note that it is the hydrogen atom in each case that actually causes the signal, but some features of each hydrogen atom's surroundings have been shown.

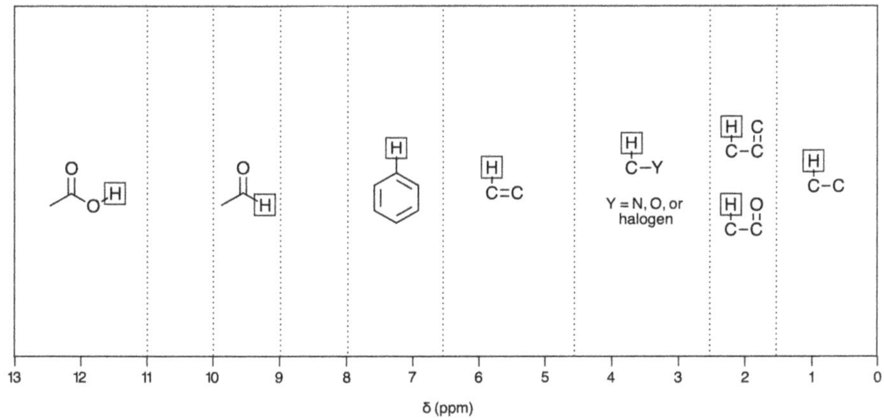

FIGURE 3.1 Correlation table for proton NMR.

There are a few features of this graphic that deserve further comment. First, it is not an exhaustive correlation table. There are other types of hydrogen atoms whose chemical shifts have not been shown. This graphic aims to convey the most commonly encountered types of hydrogens at a glance. More extensive correlation tables can easily be found online for specialized applications.

Another important note is that hydrogen-bonding protons (namely, O–H and N–H) have not been shown. It turns out their chemical shifts are highly variable, depending greatly on the concentration of the sample used to acquire the NMR spectrum. At different concentrations, hydrogen bonding will be more or less extensive, and this impacts the chemical shift. Luckily, these hydrogen-bonding protons usually reveal themselves due to the fact that their signals are typically broader than those of non-hydrogen-bonding protons. Hydrogen bonding broadens their signals at the concentrations normally used for NMR samples. It is possible to dilute an NMR sample so much that hydrogen bonding is insignificant, which then causes this broadening to disappear, but that is atypical. Most of the time, hydrogen-bonding protons will be easy to pick out because they will have broader signals.

Finally, the divisions between chemical-shift ranges are shown as dotted (not solid) lines purposefully. This is due to two factors. One is that the specific cutoffs are somewhat arbitrary and can be placed slightly differently according to preference. The other reason is that deshielding is cumulative. Deshielding is when a signal is pulled to a higher chemical-shift value by a special feature, like a functional group. For example, a hydrogen atom adjacent to a carbonyl (1.5–2.5 ppm) is deshielded relative to a hydrogen atom found in a simple alkyl group (0–1.5 ppm).

When a proton is near multiple special features, their cumulative effect may be to pull its signal to a chemical-shift value higher than we would have predicted from either feature individually. For example, consider the ester fragment shown in Figure 3.2. The indicated hydrogen atom is on a carbon adjacent to oxygen. From Figure 3.1, we would expect this to fall in the range of 2.5–4.5 ppm, likely closer to 4.5 ppm since oxygen is a fairly electronegative element. However, this oxygen atom is special in that it is connected to a carbonyl, another deshielding feature. Therefore, the cumulative effect of deshielding, by both the sp^3 oxygen and the carbonyl, may be to pull the proton's signal a bit past 4.5 ppm.

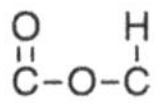

FIGURE 3.2 A proton on the carbon attached to the sp^3 oxygen of an ester.

This highlights the fact that we cannot take the regions shown in Figure 3.1 too literally. When a signal falls close to a cutoff point, we should consider it as possibly belonging to either of the adjacent regions. We will typically have context clues that help us to make the determination.

Let's practice extracting some information from proton NMR spectra using chemical shift. Consider *tert*-butyl bromide (Figure 3.3). The nine hydrogens of the three methyl groups are identical, so we expect only one signal in the proton NMR spectrum. However, the spectrum contains three signals. We can quickly resolve this discrepancy using two facts.

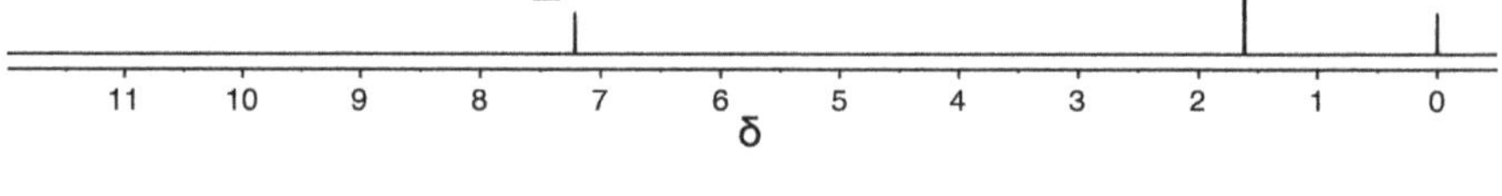

FIGURE 3.3 Proton NMR spectrum of *tert*-butyl bromide.

First, we know that TMS is used as an internal standard to mark 0 ppm, so the signal in that location is merely from TMS and is not caused by our analyte, *tert*-butyl bromide.

The second fact that we need to know deals with the NMR solvent. Typically, NMR samples are prepared by dissolving a small amount of a compound of interest (i.e., the analyte) in a suitable solvent. However, most solvents contain protons, meaning that we would then see an enormous solvent signal in the spectrum. To prevent that from happening, a deuterated solvent is used. This means that the solvent's hydrogen atoms have been replaced by deuterium, which is an isotope of hydrogen. Since deuterium (^2H) is different from hydrogen (^1H), it does not cause signals in the proton NMR. So, the solvent is mostly invisible in the proton NMR spectrum. However, it's very difficult to deuterate a substance 100%. There's always a little bit of the solvent still containing hydrogens, and this will cause a small proton NMR signal. A very common NMR solvent is deuterated chloroform ($CDCl_3$), and the trace amount of chloroform ($CHCl_3$) that remains in $CDCl_3$ will cause a small proton NMR signal at about 7.24 ppm. The bottom line is that we can simply ignore this signal as well because it is not caused by our analyte of interest.

Now that we've eliminated the signals at 0 and 7.24 ppm from consideration, we are left with one remaining signal at just over 1.5 ppm, which is due to our analyte, *tert*-butyl bromide. The protons of *tert*-butyl bromide are of the type H–C–C, so by referring to Figure 3.1, we expect their signal to fall between 0 and 1.5 ppm. The signal is just slightly past 1.5 ppm, which is due to the fact that there is a nearby halogen, which deshields these hydrogens just a bit.

Let's examine one more case study together. The structure of acetaldehyde is shown in Figure 3.4, along with the molecule's proton NMR spectrum. In the previous example, we established that the signals at 0 ppm (for the internal standard, TMS) and at 7.24 ppm (for the trace amount of chloroform in the deuterated chloroform solvent) are not of interest when analyzing a spectrum. We can effectively ignore them.

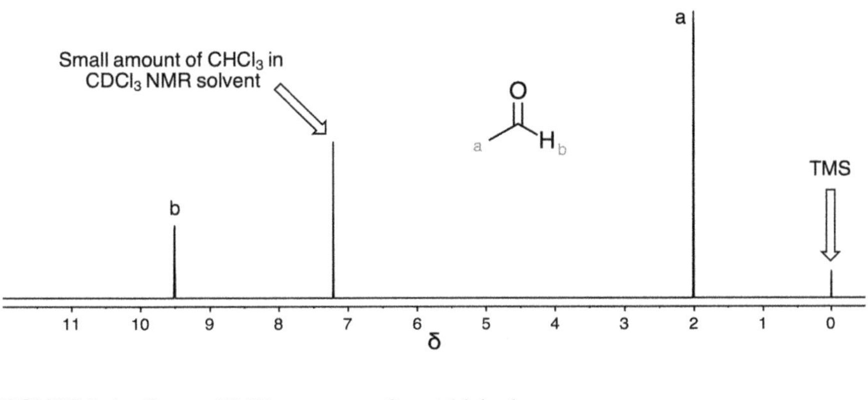

FIGURE 3.4 Proton NMR spectrum of acetaldehyde.

This leaves two signals for our consideration. One is at about 2 ppm, while the other is at approximately 9.5 ppm. Acetaldehyde contains two types of hydrogens labeled "a" and "b." The protons of type "a" are in the chemical environment H–C–C=O, which leads us to expect their signal between 1.5 and 2.5 ppm (refer to Figure 3.1). There is only one signal in that range, so we can assign it to protons "a."

The other type of hydrogen in acetaldehyde (b) is an aldehyde proton (H–C=O). Figure 3.1 tells us to expect aldehyde protons in the range of 9–10 ppm. Indeed, there is one signal in that range, so we can assign it to proton "b."

Notice how careful you must be when examining the chemical environment in which a proton finds itself. There is a significant difference in chemical shift between a proton *directly bonded* to a carbonyl (H–C=O, 9–10 ppm) and a proton on a carbon *adjacent* to a carbonyl (H–C–C=O, 1.5–2.5 ppm). Therefore, attention to detail matters, as it will impact your success in interpreting NMR spectra!

CHAPTER 3 PRACTICE PROBLEMS

1. Figure 3.5 shows the structure of 2,2-dimethyl-1-propanol, along with its proton NMR spectrum. First, assign letters to the different types of protons in 2,2-dimethyl-1-propanol. Then, label the peaks in the NMR spectrum with the letters corresponding to the protons causing these signals.

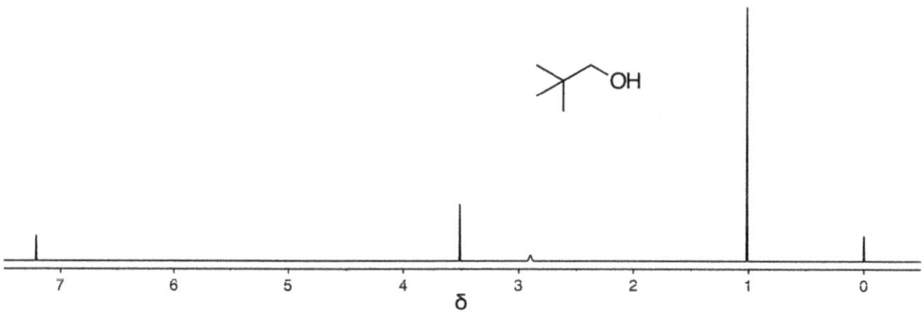

FIGURE 3.5 Proton NMR spectrum of 2,2-dimethyl-1-propanol.

2. The structure of *para*-xylene and its proton NMR spectrum are shown in Figure 3.6. Assign letters to the different types of protons in *para*-xylene, and label the peaks in the spectrum with the corresponding letters.

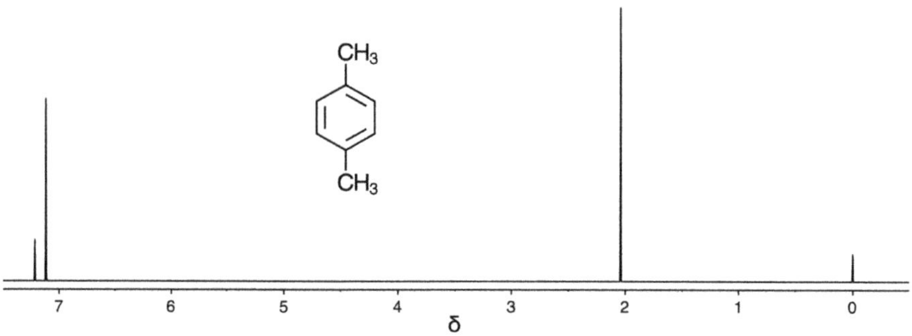

FIGURE 3.6 Proton NMR spectrum of *para*-xylene.

3. Sketch your prediction for the proton NMR spectrum of 1,3,5-tris (chloromethyl) benzene, whose structure is shown in Figure 3.7.

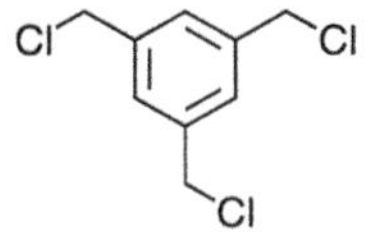

FIGURE 3.7 1,3,5-Tris(chloromethyl)benzene.

4. Draw the expected proton NMR spectrum of *tert*-butyl methyl ether (Figure 3.8).

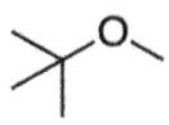

FIGURE 3.8 *tert*-Butyl methyl ether.

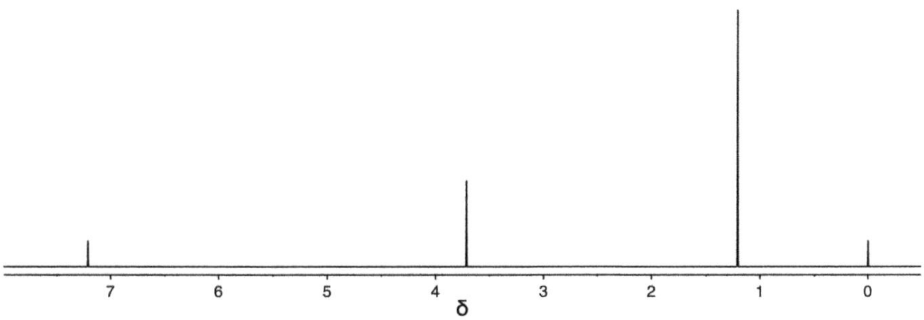

FIGURE 3.9 Proton NMR of an unknown ester ($C_6H_{12}O_2$).

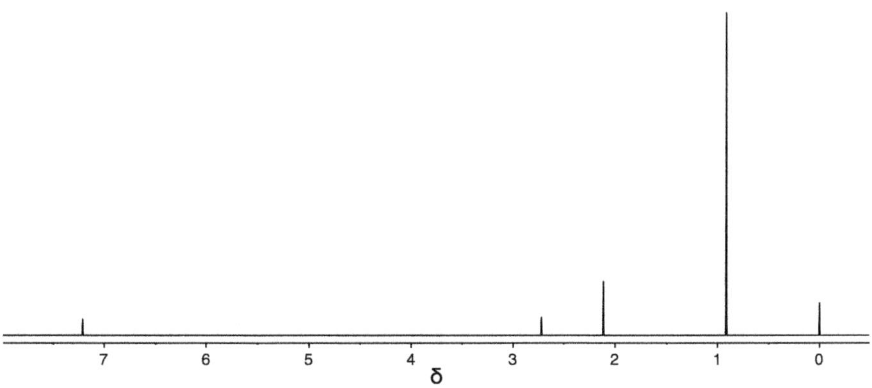

FIGURE 3.10 Proton NMR of an unknown ketone ($C_{11}H_{22}O$).

5. Draw the structure of an ester having the molecular formula $C_6H_{12}O_2$ that is consistent with the proton NMR spectrum in Figure 3.9.
6. Identify the structure of a ketone having the molecular formula $C_{11}H_{22}O$ that is consistent with the proton NMR spectrum shown in Figure 3.10.

SOLUTIONS TO CHAPTER 3 PRACTICE PROBLEMS

1. 2,2-Dimethyl-1-propanol has three types of protons. The nine protons of the three methyl groups (a) are all equivalent. Be careful because the quaternary carbon of the *tert*-butyl group does not possess any hydrogens, so there will be no label for that location. The methylene (CH_2) group is a second type of proton (b). Finally, the hydrogen of the hydroxyl (OH) group is a third type of proton (c).

Now, we must match these labels with the corresponding proton NMR signals (Figure 3.11). The first thing for us to do is to tag the peaks at 0 and 7.24 ppm as TMS and residual $CHCl_3$, respectively. These two peaks are irrelevant to our interpretation of the spectrum.

Then, only three signals remain to match our three types of protons. Protons of type "a" are in the chemical environment H–C–C, so our correlation table (Figure 3.1) tells us to expect them in the range 0–1.5 ppm. There is only one signal in that region, so we can assign it the label "a."

As we move on to the assignment of the remaining two signals as "b" and "c," we encounter a small challenge. Both of these signals fall in the chemical-shift range 2.5–4.5 ppm, so how will we distinguish them? The answer lies in one of the caveats stated about Figure 3.1. We noted that hydrogen-bonding protons were not shown in that graphic because their chemical shifts vary greatly depending upon concentration. However, we also noted that their signals are distinctive due to broadening resulting from the hydrogen bonding. The signal just below 3 ppm is significantly broader than the other signals in the NMR spectrum. Therefore, we can immediately identify it as the only proton capable of hydrogen bonding (c).

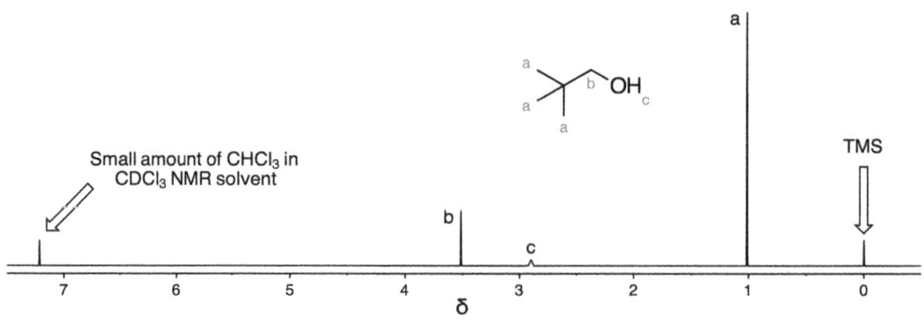

FIGURE 3.11 Labeled proton NMR spectrum of 2,2-dimethyl-1-propanol.

With that assignment made, only one signal remains at about 3.5 ppm. This signal can be assigned to protons of type "b," which are in the chemical environment H–C–O and therefore appear between 2.5 and 4.5 ppm.

Here's an important note. If you looked at protons "b" and thought they should be categorized as being in the chemical environment H–C–C (because you followed a path to the left in the structure), you'll want to be aware that we always look for the most "special" or distinguishing feature when we assign a chemical environment for the determination of chemical shift. In other words, when looking at a proton of type "b," it is clearly connected to a carbon (H–C). When you move one bond further from the proton and see two choices (H–C–C vs. H–C–O), you identify the oxygen as a more special or unusual structural feature and select H–C–O as the best description of this proton's chemical environment. Remember that carbons, hydrogens, and single bonds are the norm for organic molecules. Anything else is more special or distinguishing.

2. The aromatic compound *para*-xylene has two axes of symmetry that cut through the molecule both horizontally and vertically. As a result, it contains only two different types of protons (Figure 3.12): those of the methyl groups (a) and the four aromatic hydrogens on the ring, which are all identical to one another (b).

Remember that the peaks at 0 and 7.24 ppm denote TMS and trace $CHCl_3$, respectively, and are therefore of no interest when assigning the signals.

The protons of type "a" are in the chemical environment H–C–C=C, so the correlation table (Figure 3.1) leads us to expect them between 1.5 and 2.5 ppm, allowing us to assign the signal in that region.

In contrast, the protons of type "b" are in the aromatic chemical environment,

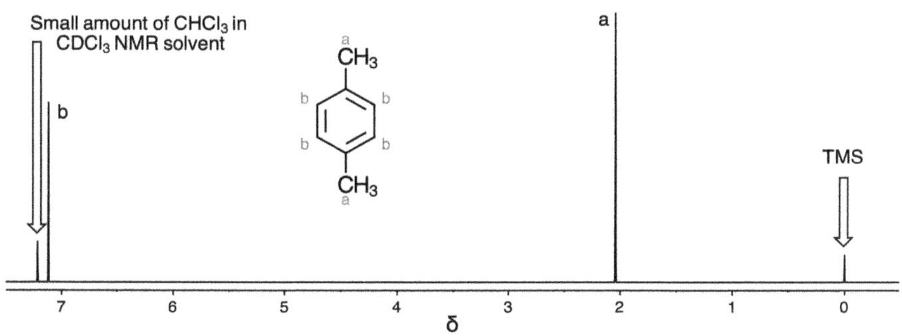

FIGURE 3.12 Labeled proton NMR spectrum of *para*-xylene.

so the correlation table indicates that we should find them between 6.5 and 8 ppm. Consequently, we can assign the only remaining signal in that region once we've decided to disregard the solvent signal.

Again, note how carefully we must examine the chemical environment. There's a big chemical shift difference between protons *on* an aromatic ring and protons on a carbon *adjacent to* an aromatic ring.

3. When predicting a proton NMR spectrum from a structure, we must first decide how many signals we should see. This means that we must identify the number of types of protons in the structure. 1,3,5-Tris(chloromethyl) benzene has only two types of protons (Figure 3.13): those of the three methylene groups (a), which are all identical to one another given the molecule's high degree of symmetry, and the aromatic hydrogens (b), which are also all identical to one another because of symmetry.

When sketching the proton NMR spectrum, we can first add the TMS signal at 0 ppm and the solvent signal at 7.24 ppm. Then, in order to place the signals for protons "a" and "b" correctly, we'll need to classify their chemical environment. The protons labeled "a" are of the type H–C–Cl, which leads us to expect them in the range of 2.5–4.5 ppm. Given that these same protons are also on a carbon

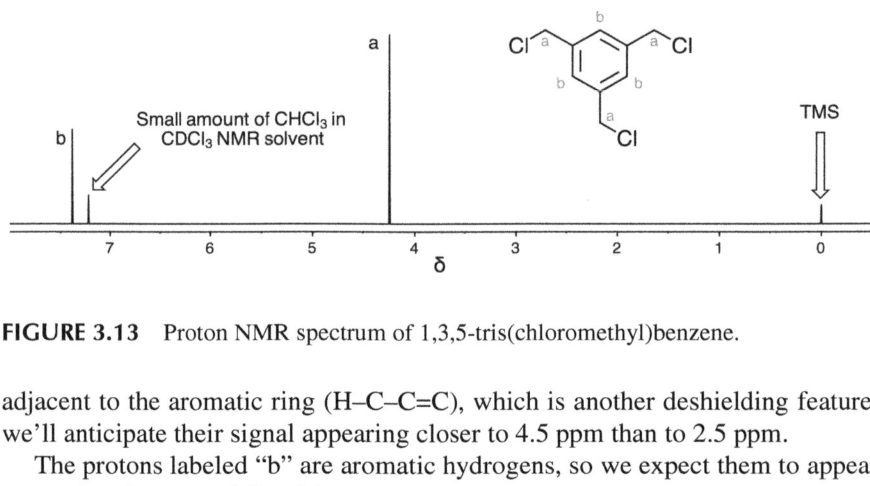

FIGURE 3.13 Proton NMR spectrum of 1,3,5-tris(chloromethyl)benzene.

adjacent to the aromatic ring (H–C–C=C), which is another deshielding feature, we'll anticipate their signal appearing closer to 4.5 ppm than to 2.5 ppm.

The protons labeled "b" are aromatic hydrogens, so we expect them to appear somewhere between 6.5 and 8 ppm.

Naturally, some small variation in where you place the signals within their assigned regions is acceptable. Also, for the time being, you need not worry about the size of the signals. We'll address that in the next chapter.

4. Our first step should be to ascertain the number of types of hydrogens in *tert*-butyl methyl ether, which will tell us how many signals to place in its proton NMR. There are only two types of hydrogens (Figure 3.14): those of the three equivalent methyls of the *tert*-butyl group (a) and those of the methyl group directly bonded to oxygen (b).

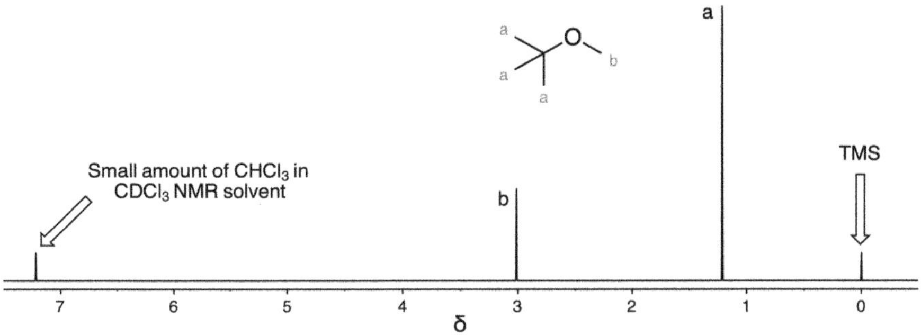

FIGURE 3.14 Proton NMR spectrum of *tert*-butyl methyl ether.

As we begin to draw the spectrum, we can insert the signals for TMS at 0 ppm and for the solvent at 7.24 ppm. Next, we'll classify each type of proton in *tert*-butyl methyl ether to determine the chemical-shift range in which it should appear. The protons labeled "a" are of the type H–C–C and therefore appear in the range of 0–1.5 ppm according to Figure 3.1. The protons labeled "b" are of the type

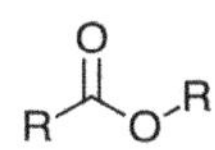

FIGURE 3.15 An ester.

H–C–O, so our correlation table tells us to expect them between 2.5 and 4.5 ppm.

Once again, for the time being, you need not worry about the size of the signals. We'll address that in the next chapter.

5. The question tells us that the unknown molecule is an ester (Figure 3.15). The functional group alone accounts for one carbon (the carbonyl carbon) and two oxygen atoms. If we subtract them from the original molecular

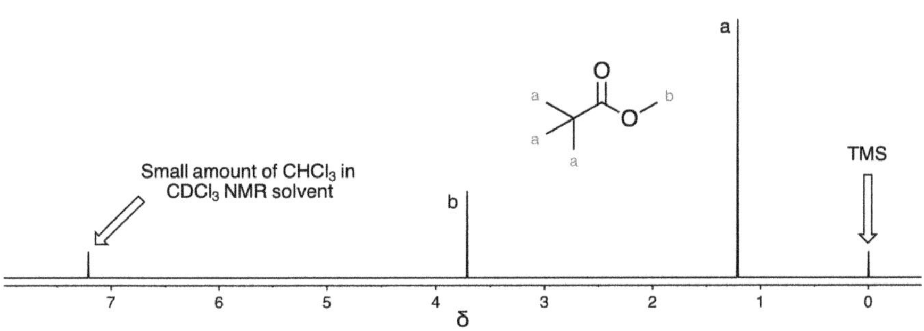

FIGURE 3.16 Proton NMR spectrum of methyl 2,2-dimethylpropionate.

formula ($C_6H_{12}O_2$), there are five carbons and twelve hydrogens left to place in the ester's two R groups.

The proton NMR spectrum for the unknown has signals at 0 ppm (TMS) and 7.24 ppm (solvent) that we should disregard (Figure 3.16). Beyond that, it has only two signals. One is in the range 0–1.5 ppm, which corresponds to protons in the chemical environment H–C–C. The other is in the range 2.5–4.5 ppm, which indicates protons of the type H–C–Y, where Y = N, O, or a halogen. Given that the molecular formula contains no nitrogen or halogen, we can say that these protons are of the type H–C–O. If we place a methyl group (b) on the sp^3 hybridized oxygen atom of the ester, we can account for this signal.

After placing the methyl group "b," we are left with four carbons and nine hydrogens. In order for all of the hydrogen atoms (a) to be chemically equivalent,

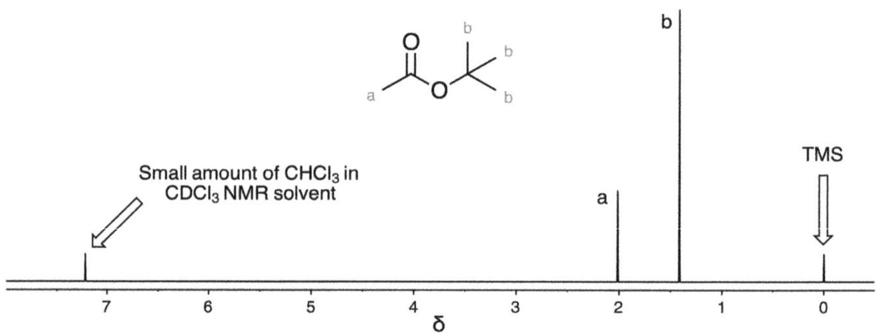

FIGURE 3.17 Proton NMR spectrum of *tert*-butyl acetate.

we'll have to use a *tert*-butyl group. The compound that results is known as methyl 2,2-dimethylpropionate. Once again, don't worry about the relative sizes of the signals just yet; that will be addressed in the next chapter.

Notice that we could not swap the methyl and *tert*-butyl groups of the ester. If

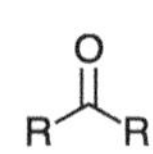

FIGURE 3.18 A ketone.

we did that, the resulting molecule is *tert*-butyl acetate, and its proton NMR spectrum is different from the one we were given (Figure 3.17). *tert*-Butyl acetate also contains only two types of protons; however, they are different types: H–C–C=O (1.5–2.5 ppm) for the protons labeled "a" and H–C–C (0–1.5 ppm) for the protons labeled "b." As a result, this molecule does *not* yield any signals in the range of

2.5–4.5 ppm, as our unknown did. So, *tert*-butyl acetate would *not* be a correct answer to the question.

6. The question tells us that the unknown molecule is a ketone (Figure 3.18). The ketone's carbonyl (C=O) accounts for one carbon and the oxygen. This leaves ten carbons and twenty-two hydrogens to place in the ketone's two R groups.

As we turn our attention to the spectrum (Figure 3.19), remember to ignore the TMS signal at 0 ppm and the solvent signal at 7.24 ppm. This leaves only three signals, meaning that the twenty-two hydrogens are divided into a mere three types. This implies some symmetry.

We have one signal in the range of 2.5–4.5 ppm (b), which typically indicates a proton in the chemical environment H–C–Y, where Y = N, O, or a halogen. However, in this case, there's only one of these atoms, and that oxygen is already accounted for in the ketone's carbonyl. This oxygen cannot have any more bonds, and the ketone's carbonyl carbon cannot bear any hydrogens, or it would be an aldehyde rather than a ketone.

That seems to be a bit of a conundrum. However, remember one of our caveats about the correlation table:

> we cannot take the regions shown in Figure 3.1 too literally. When a signal falls close to a cutoff point, we should consider it as possibly belonging to either adjacent region. We will typically have context clues that help us to make the determination.

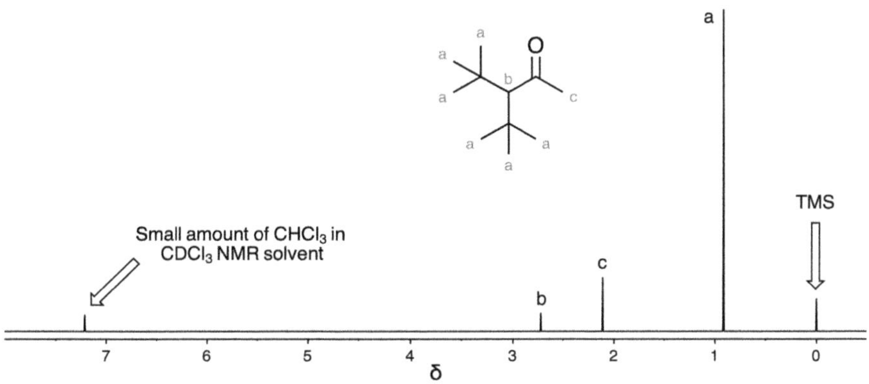

FIGURE 3.19 Proton NMR spectrum of 3-*tert*-butyl-4,4-dimethyl-2-pentanone.

Here, the context clue is that it is impossible, given the constraints of this problem, to have a proton of type H–C–Y in this molecule. Therefore, the signal just past 2.5 ppm likely belongs in the other adjacent region of the correlation table: H–C–C=O (1.5–2.5 ppm). This type of hydrogen is much more attainable in this question. There's also a second signal in this range of 1.5–2.5 ppm (c), so we must have two different types of protons on carbons adjacent to the carbonyl. This will

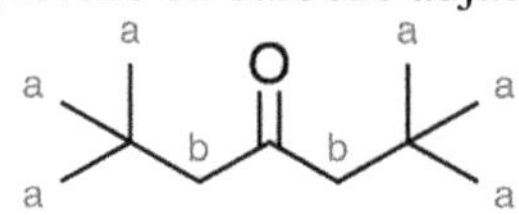

FIGURE 3.20 2,2,6,6-Tetramethyl-4-heptanone.

mean that the two R groups of the ketone differ in structure. By placing positions "b" and "c" adjacent to the carbonyl, we have accounted for three out of eleven carbons. Eight carbons remain.

We have only one remaining type of hydrogen atom (a). All the hydrogens stemming from the remaining carbons must be chemically equivalent, or else they would cause more than one signal. In order to make all the hydrogens stemming from the remaining carbons equivalent, we are forced to use two *tert*-butyl groups, and they must stem from the same point of origin. Therefore, the unknown ketone is shown to be 3-*tert*-butyl-4,4-dimethyl-2-pentanone.

Notice that you could *not* place one *tert*-butyl group on each side of the molecule (as shown in Figure 3.20) because that would result in a molecule with too much symmetry for this problem. It would have only two types of protons (a and b) and would therefore give only two proton NMR signals, which does not match what we see in the spectrum. Therefore, this would be an *incorrect* answer.

4 Proton NMR Part III
Integration

It probably doesn't surprise you that the size of a nuclear magnetic resonance (NMR) signal is proportional to the number of protons causing that signal. You may have even noticed in spectra from previous chapters that some signals were larger than others and made this connection on your own.

Consider the proton NMR spectrum of acetaldehyde (Figure 4.1). The signal labeled "a" (of the type H–C–C=O; 1.5–2.5 ppm according to Figure 3.1) stems from a methyl (CH_3) group and is therefore due to three hydrogen atoms, while the signal labeled "b" is caused by the single aldehyde proton (9–10 ppm). Therefore, the signal for protons "a" is three times as big as the signal for proton "b."

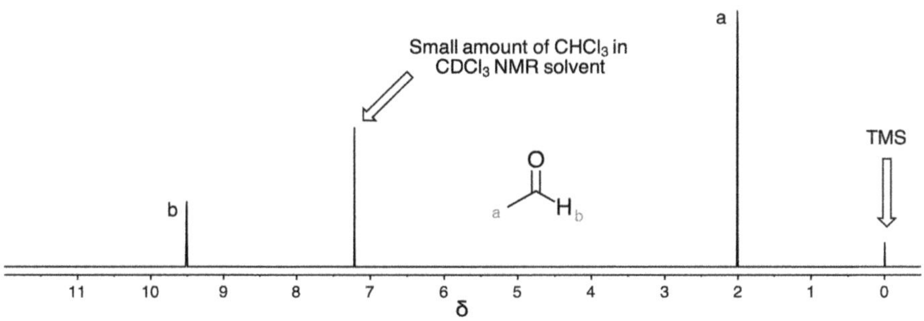

FIGURE 4.1 Proton NMR spectrum of acetaldehyde.

However, we have to be careful how we measure a signal's size. Height seems like an easy choice, but signals can have vastly different shapes. The shapes of proton NMR signals will be addressed more fully in the next chapter, but for the time being, consider the proton NMR spectrum of 2-butanone (Figure 4.2).

There are three types of hydrogen. Type "a" is H–C–C and therefore appears between 0 and 1.5 ppm according to our correlation table (Figure 3.1). The protons labeled "b" and "c" are all of the type H–C–C=O and therefore fall between 1.5 and 2.5 ppm. We'll need some additional information to distinguish them from one another. One such piece of information is the size of the signals (this chapter), and another is the shape of the signals (the next chapter).

As noted above, we have to be careful how we compare the sizes of signals. Signals "a" and "c" are both due to methyl (CH_3) groups and are therefore caused

DOI: 10.1201/9781003783329-4

by the same number of hydrogen atoms (three). These signals have different heights, though, and that is due to their differing shapes. One signal consists of a single spike, while the other is made up of a few spikes.

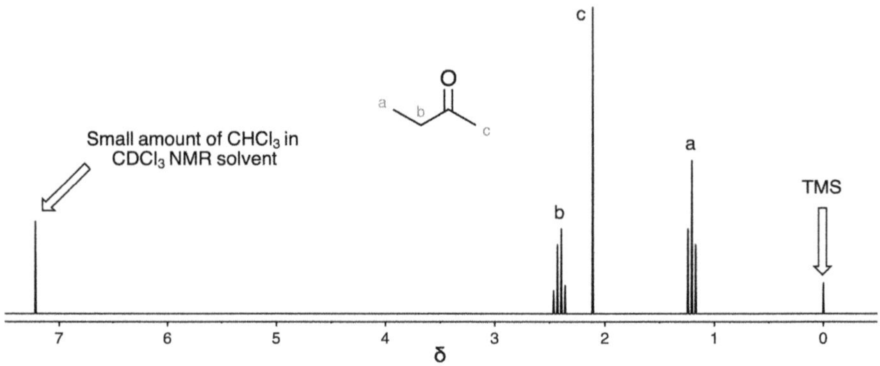

FIGURE 4.2 Proton NMR spectrum of 2-butanone.

When signals have different shapes, which they often do, it is not fair to compare their heights. That is to say, a tall, skinny peak may be of the same "size" as a shorter, squatter signal, though they differ in height. So, to be fair, we compare their integrations. Integration is the area under the curve caused by each signal. You may be familiar with this term if you've had calculus, though we will not be calculating any integrals ourselves. Software packages compute the area under the curve for us, and these integrals are printed on the spectrum. Note that the location of the integral can vary depending on the software package used. Throughout this book, the integration values will be expressed as a specific number of hydrogens (3H, 2H, 1H, etc.) and will be displayed near the signals but above the x-axis, though some software shows them below the x-axis.

When we re-examine the proton NMR spectrum of 2-butanone with integrals (Figure 4.3), we are able to distinguish between signals "b" and "c," which were difficult to differentiate previously because they fell within the same chemical-shift range. Signal "b" integrates for only two hydrogens, so it must correlate with the methylene (CH_2) group. Signal "c," on the other hand, integrates for three hydrogens, so it must be the methyl group that is adjacent to the carbonyl.

There is a second way that integration data can be presented, which is as an integral trace. The integral trace is the dotted line in Figure 4.4. An integral trace is a line inserted by the software package that is horizontal as long as there is no area underneath the curve (i.e., along the spectrum's baseline). When the integral trace reaches a signal, there is a vertical jump that is proportional to the area under the curve of that peak.

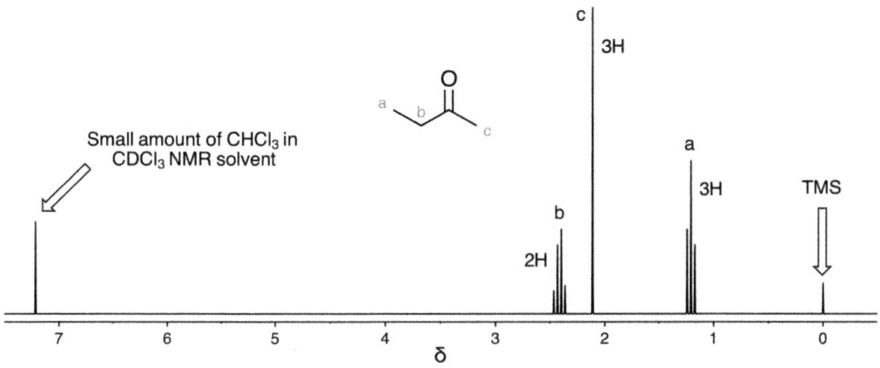

FIGURE 4.3 Proton NMR spectrum of 2-butanone with integration values.

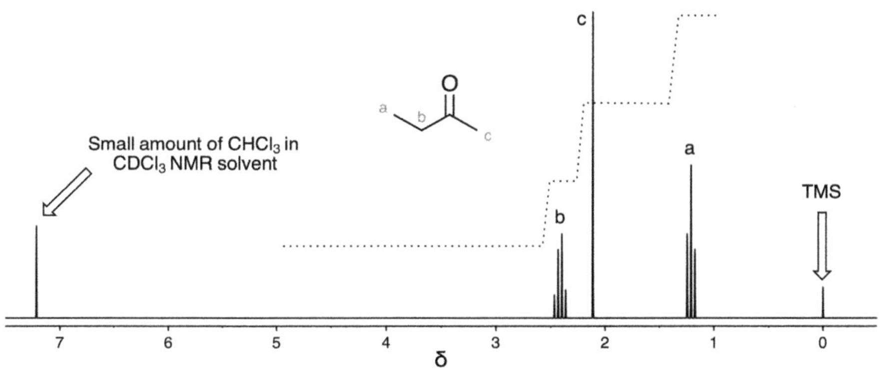

FIGURE 4.4 Proton NMR spectrum of 2-butanone with an integral trace.

We can convert an integral trace into integration values as follows. First, you'll measure the vertical jumps from plateau to plateau at each peak (Figure 4.5). These are shown below from left to right as "distance b" for peak "b," "distance c" for peak "c," and "distance a" for peak "a." Specific measurements have not been included here because you may be reading this in a hardcopy textbook or on a screen, and the distances will depend on how you are viewing the spectrum. For your specific viewing experience, simply use a ruler to measure each bracketed distance.

Then, we have to normalize these values by dividing them all by the smallest one. In other words, the specific distances depend on exactly how you are viewing this diagram, so raw distance measurements mean little. Instead, we want to develop a ratio of these distances to one another.

Regardless of how you are reading this and what your specific measurements are, distance b is the smallest, so we divide all three measurements by distance b. This will give the following ratios:

For peak "b," $\frac{\text{distance b}}{\text{distance b}} = 1.$

For peak "c," $\frac{\text{distance c}}{\text{distance b}} = 1.5.$

For peak "a," $\frac{\text{distance a}}{\text{distance b}} = 1.5.$

Since we cannot have half protons, multiply *all* values by two to attain the ratio 2:3:3 for protons b:c:a. These are now the same integration values (2H:3H:3H) that were provided in Figure 4.3. We simply arrived at them in a different way. Whether you see integral values as numbers printed near each peak or as an integral trace simply depends on the software package you use when acquiring NMR spectra in the laboratory, so you need to be able to work with both presentation methods.

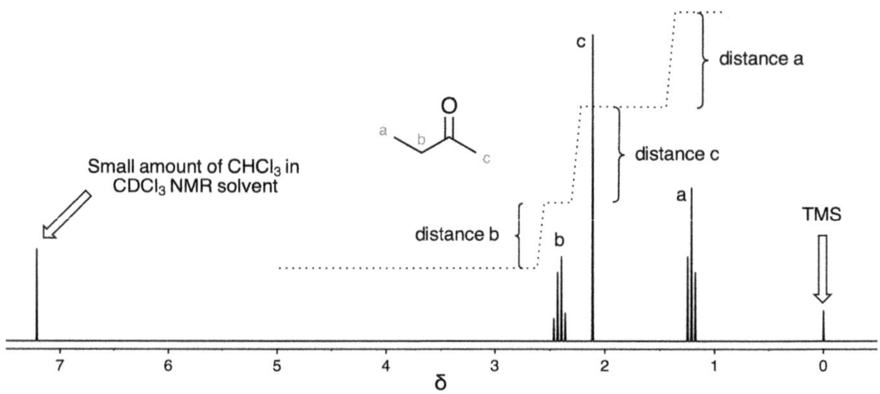

FIGURE 4.5 Proton NMR spectrum of 2-butanone with integral-trace measurements.

Let's try one more example together in which we highlight the value of integration in making peak assignments. An unlabeled proton NMR of isobutyl alcohol is given in Figure 4.6. By examining the structure of isobutyl alcohol,

FIGURE 4.6 Proton NMR spectrum of isobutyl alcohol with integration values.

we determine that four signals should exist: one for the two identical methyl (CH_3) groups (a) stemming from the same carbon, one for the methine (CH) group (b), one for the methylene (CH_2) group (c), and one for the hydroxyl (OH) proton (d).

Some of the signals may be assigned on the basis of chemical shift alone, though integration will be helpful in confirming our choices. For example, the hydrogens of the methyl groups are of the type H–C–C and should therefore appear between 0 and 1.5 ppm. There's only one signal in that region (at about 1 ppm), so we can easily assign it to protons "a." Note that the integration is tremendously helpful in validating this choice. The two methyl groups account for six hydrogens, and the integration of that peak is six. While the shape of the signal (two spikes, rather than just one) is something different from what we've seen thus far, don't worry about that at this point. The next chapter will explain peak shape in detail.

There is a signal just below 2 ppm that is initially a bit mysterious. The chemical-shift range 1.5–2.5 ppm usually denotes protons on a carbon adjacent to a carbon-oxygen or carbon-carbon double bond (H–C–C=O or H–C–C=C). Neither of those structural features appears in this molecule, so why is there a signal in this region? The methine (CH) group (b) is formally of the type H–C–C, so you might have expected it to appear between 0 and 1.5 ppm as well. However, this proton is getting closer to the alcohol functional group. As a result, it's not a garden-variety H–C–C. The relatively close, deshielding oxygen atom pulls it a bit further downfield, slightly past the normal 1.5 ppm cutoff for the H–C–C range. Further support for this assignment comes from the fact that the signal integrates for one hydrogen, which matches the expected integration for "b." True, this signal has an unusual shape; we'll reveal the reason for that as we continue our journey through NMR in the next chapter.

The next signal at just below 3 ppm also integrates for one hydrogen atom. However, it is a broad signal relative to the other sharp spikes we see in the spectrum. In Chapter 3, we established that hydrogen-bonding protons, such as the hydroxyl (OH) group's proton, will have variable chemical shifts but reveal themselves through broadened signals.

Lastly, at 3.5 ppm, we see a signal that should correspond to protons of the type H–C–O. Protons "c" fall into this category, and since there are two protons in this methylene group, the integration of two is as expected.

Now that we've made all of the assignments, Figure 4.7 summarizes our findings.

One last note about integral values is that they are merely ratios, not necessarily absolute numbers of hydrogen atoms. When you are acquiring an NMR spectrum, the software typically does not know what molecule you are analyzing. Therefore, it has no idea how many hydrogen atoms there should be. It is merely providing you with the ratio of area under the curve for each signal, and you usually receive that ratio in its most reduced form. For example, if you have signals caused by three hydrogens and nine hydrogens, the integral values

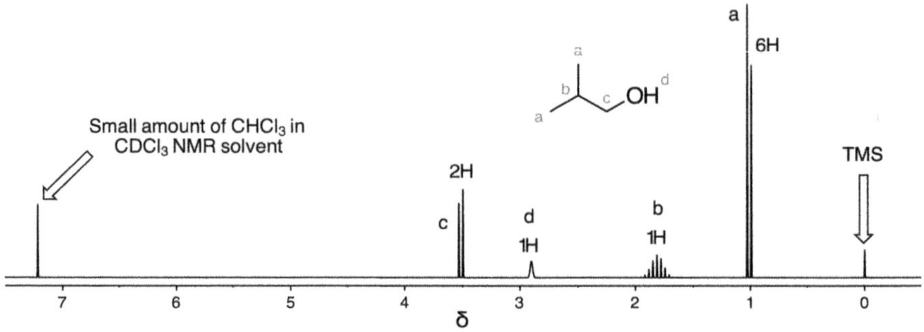

FIGURE 4.7 Labeled proton NMR spectrum of isobutyl alcohol.

may appear as 1:3, which is a reduced form of the ratio 3:9. If you know what molecule you are analyzing, the software will typically have an option that allows you to adjust the ratio accordingly (e.g., changing a ratio of 1:3 to its equivalent of 3:9).

CHAPTER 4 PRACTICE PROBLEMS

1. Revisit each of the previous chapter's problems (Chapter 3, Problems 1–6) and add integration values to each of the peaks that you previously assigned.
2. Assign the peaks in the proton NMR spectrum of 1-*tert*-butoxy-2-methylpropane (Figure 4.8).

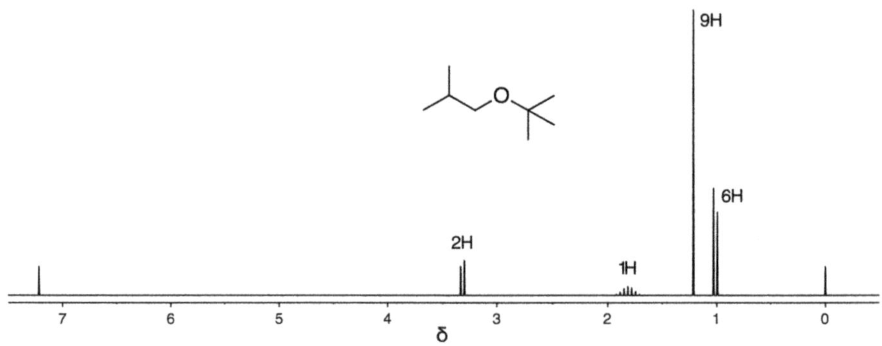

FIGURE 4.8 Proton NMR spectrum of 1-*tert*-butoxy-2-methylpropane.

3. Assign the peaks in the proton NMR spectrum of ethyl 2-methylpropionate (Figure 4.9).

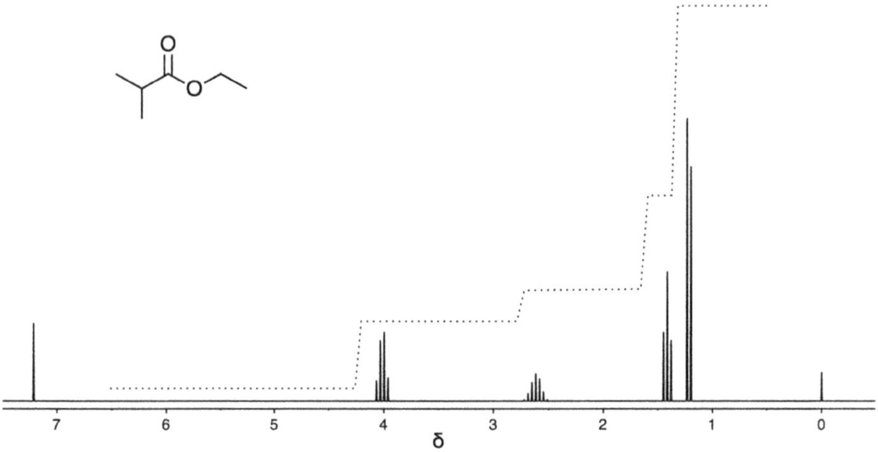

FIGURE 4.9 Proton NMR spectrum of ethyl 2-methylpropionate.

4. Determine the structure of an alkyl halide with the molecular formula $C_6H_{13}Br$ that is consistent with the proton NMR spectrum in Figure 4.10.

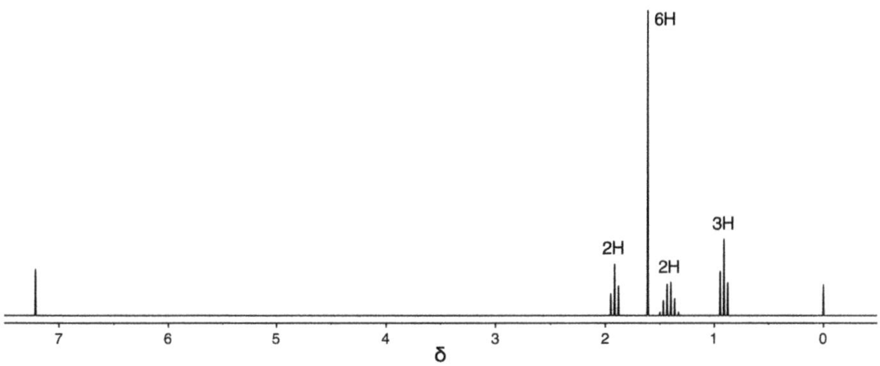

FIGURE 4.10 Proton NMR spectrum of an unknown alkyl bromide ($C_6H_{13}Br$).

SOLUTIONS TO CHAPTER 4 PRACTICE PROBLEMS

1.

a. Chapter 3, Problem 1
In the spectrum of 2,2-dimethyl-1-propanol (Figure 4.11), there are nine protons of type "a," two of type "b," and one of type "c." These integration values are placed near each peak, and the integration ratio of 9:2:1 cannot be further reduced.

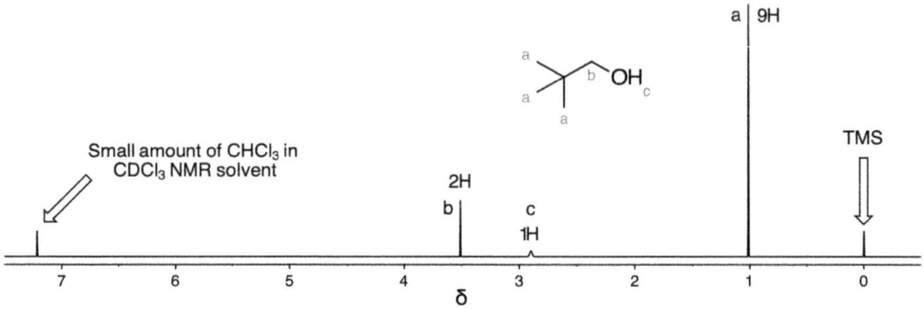

FIGURE 4.11 Proton NMR spectrum of 2,2-dimethyl-1-propanol with integration values.

b. Chapter 3, Problem 2
In the proton NMR spectrum of *para*-xylene (Figure 4.12), there are six protons of type "a" and four protons of type "b." These values are placed near each signal. Note, though, that this ratio of 6:4 may also appear in its reduced form of 3:2. That would be equally correct.

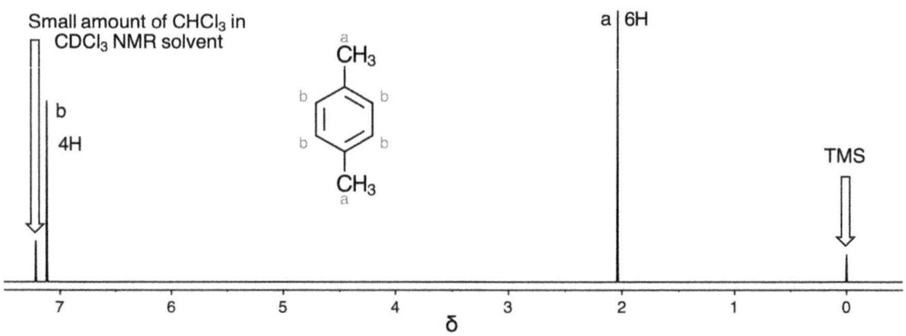

FIGURE 4.12 Proton NMR spectrum of *para*-xylene with integration values.

c. Chapter 3, Problem 3
In the proton NMR of 1,3,5-tris(chloromethyl)benzene (Figure 4.13), there are two signals, "a" and "b," caused by six and three protons, respectively. This ratio of 6:3 could also appear in its reduced form of 2:1.

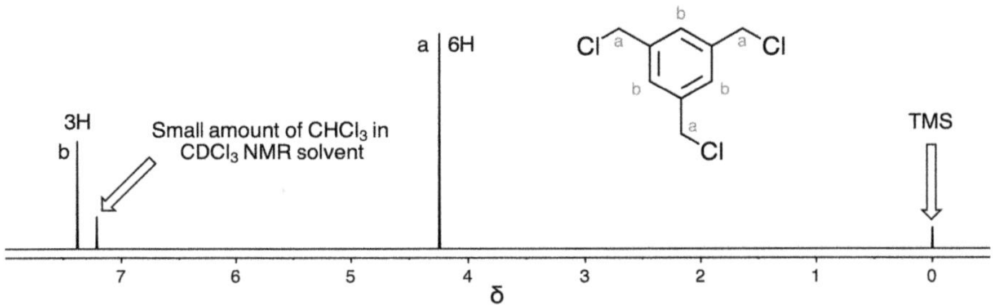

FIGURE 4.13 Proton NMR spectrum of 1,3,5-tris(chloromethyl)benzene with integration values.

d. Chapter 3, Problem 4
The proton NMR spectrum of *tert*-butyl methyl ether (Figure 4.14) shows two types of hydrogen atoms labeled "a" and "b." There are nine hydrogens of type "a" and three protons of type "b." Those values are placed near each signal. However, you may also see the ratio of 9:3 in its reduced form of 3:1, which is equally correct.

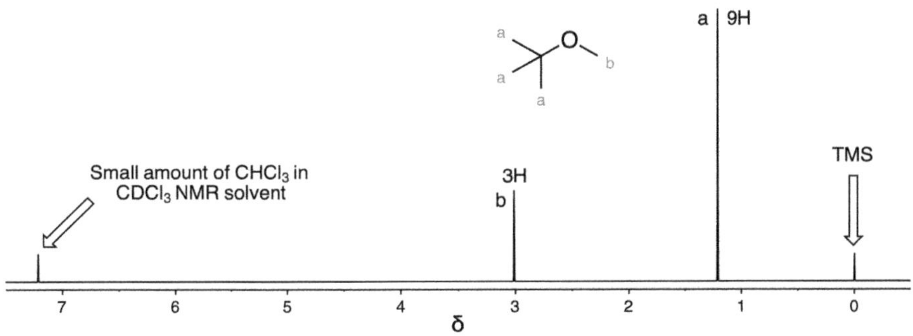

FIGURE 4.14 Proton NMR spectrum of *tert*-butyl methyl ether with integration values.

e. Chapter 3, Problem 5
The proton NMR of methyl 2,2-dimethylpropionate (Figure 4.15) also has two signals, "a" and "b," caused by nine and three protons, respectively. Note that the ratio of 9:3 could also be expressed simply as 3:1.

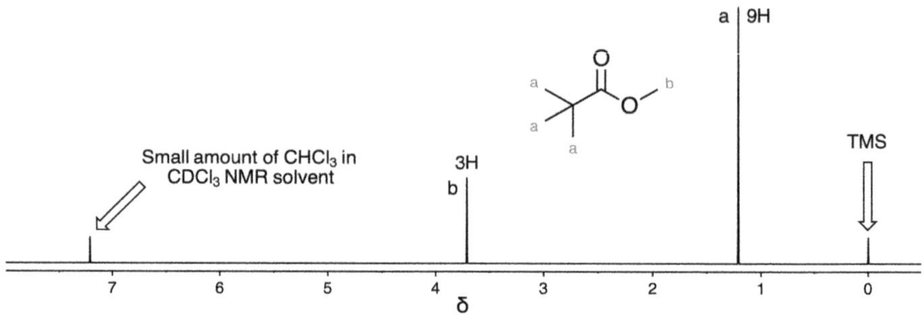

FIGURE 4.15 Proton NMR spectrum of methyl 2,2-dimethylpropionate with integration values.

f. Chapter 3, Problem 6
   The proton NMR spectrum for 3-*tert*-butyl-4,4-dimethyl-2-pentanone (Figure 4.16) contains three signals (a, b, and c). There are eighteen hydrogens of type "a," one hydrogen of type "b," and three hydrogens of type "c." The integration ratio of 18:1:3 cannot be further reduced.

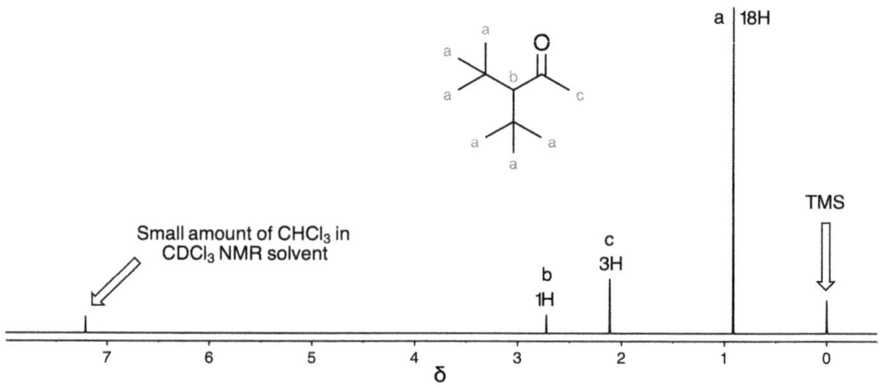

FIGURE 4.16 Proton NMR spectrum of 3-*tert*-butyl-4,4-dimethyl-2-pentanone with integration values.

2. We should begin by labeling the different types of hydrogens in 1-*tert*-butoxy-2-methylpropane with letters (Figure 4.17). There are four types of hydrogen atoms in this molecule, so we expect four signals. Once we label the TMS and solvent signals, there are four remaining signals with integration values. Some of these signals have unusual shapes, and that will be addressed in the next chapter. For now, we need only concern ourselves with the chemical shift and integration values.

In this molecule, each signal is caused by a unique number of hydrogens, making the assignments extraordinarily straightforward. There are six protons of type "a," so the corresponding signal is the one integrating for six hydrogens. There is only one proton of type "b," so that label goes with the sole signal having an integration of one hydrogen. The methylene group "c" has two hydrogens, so the corresponding signal has an integration value of two. Finally, the three equivalent methyl groups "d" have nine hydrogens and therefore correlate with the signal integrating for nine.

Chemical shift data reinforce our conclusions. Protons of both types "a" and "d" are in the chemical environment H–C–C and appear between 0 and 1.5 ppm as expected. The methine proton (b) is also of the type H–C–C but is pulled slightly further downfield. Methine protons tend to be more deshielded than comparable methylene and methyl groups. Also, this methine is a bit closer to the oxygen atom than the protons labeled "a." Finally, the methylene protons (c) are of the type H–C–O and consequently appear between 2.5 and 4.5 ppm as anticipated.

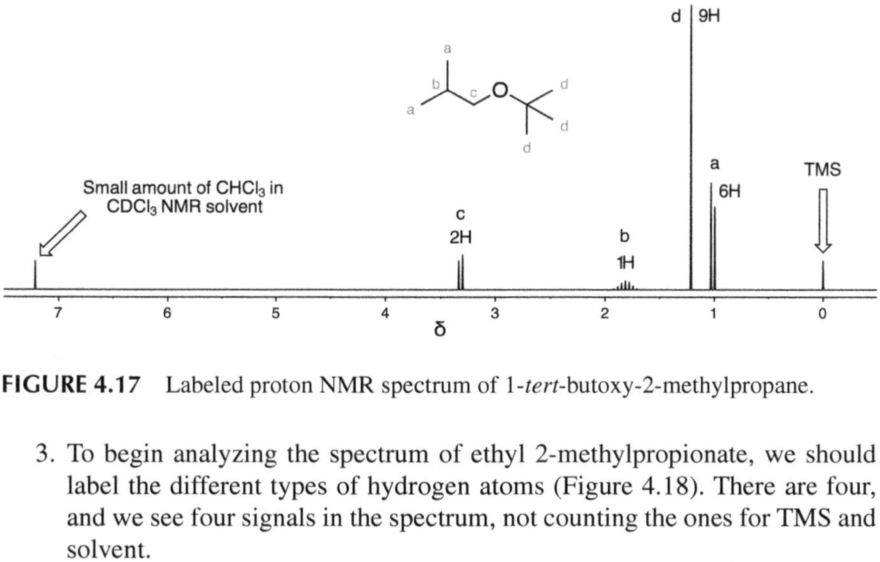

FIGURE 4.17 Labeled proton NMR spectrum of 1-*tert*-butoxy-2-methylpropane.

3. To begin analyzing the spectrum of ethyl 2-methylpropionate, we should label the different types of hydrogen atoms (Figure 4.18). There are four, and we see four signals in the spectrum, not counting the ones for TMS and solvent.

A good second step would be to recognize that an integral trace (the dotted line) has been provided, and we need to measure each increase in height from plateau to plateau. Those specific distances will be different whether you are reading this book in hardcopy form or on your screen, but whatever distances you measure with a ruler, divide them all by the smallest one (distance b). This will yield an integration ratio of 2:1:3:6 for the distances labeled c:b:d:a, respectively.

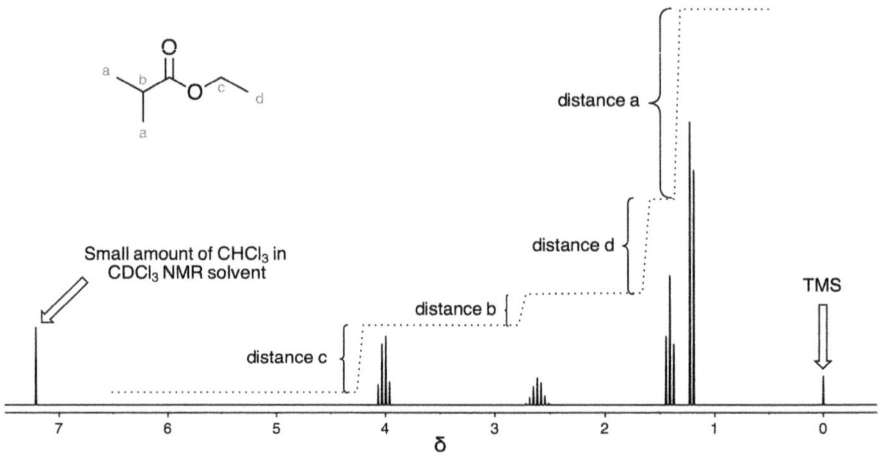

FIGURE 4.18 Proton NMR spectrum of ethyl 2-methylpropionate with integral-trace measurements.

Now that we have the integration ratio, we can replace the integral trace with the integration values (Figure 4.19). Since each signal is caused by a unique number of hydrogens, we can make assignments based on the integration values alone. There are six protons of type "a," one of type "b," two of type "c," and three of type "d."

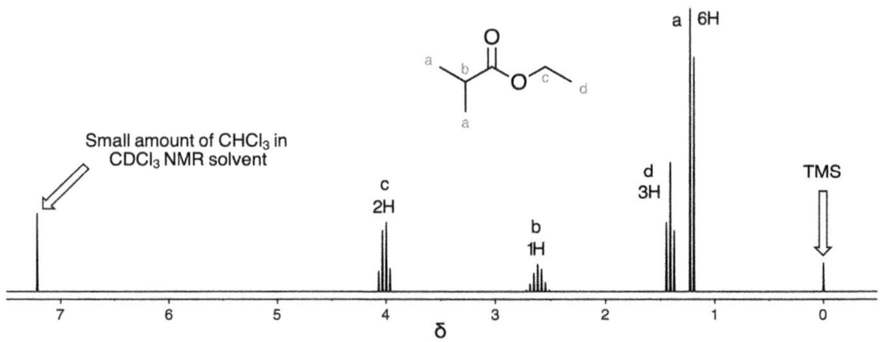

FIGURE 4.19 Labeled proton NMR spectrum of ethyl 2-methylpropionate.

Chemical-shift values reinforce our conclusions. Protons "a" and "d" are of the type H–C–C, so they appear between 0 and 1.5 ppm. Protons "d" are closer to an electronegative oxygen atom, so they are slightly more deshielded than "a." The methine proton (b) is in the chemical environment H–C–C=O, so we expect it in the range 1.5–2.5 ppm. It is just slightly past 2.5 ppm because the carbonyl carbon is bonded to an additional electronegative oxygen atom, causing cumulative deshielding. Finally, the methylene protons (c) are of the type H–C–O and therefore appear between 2.5 and 4.5 ppm.

4. The four signals have integration values showing one unique methyl group (3H), two distinct methylenes (2H each), and two methyl groups (6H) that are different from the first methyl but identical to one another. If these methyls are identical because they stem from the same carbon, then these fragments account for all of the carbon and hydrogen atoms. Remembering that there is also a bromine atom, we have all the fragments in hand. There are only two possible ways to assemble these fragments while preserving exactly four types of hydrogens: 2-bromo-2-methylpentane or 1-bromo-2,2-dimethylbutane (Figure 4.20).

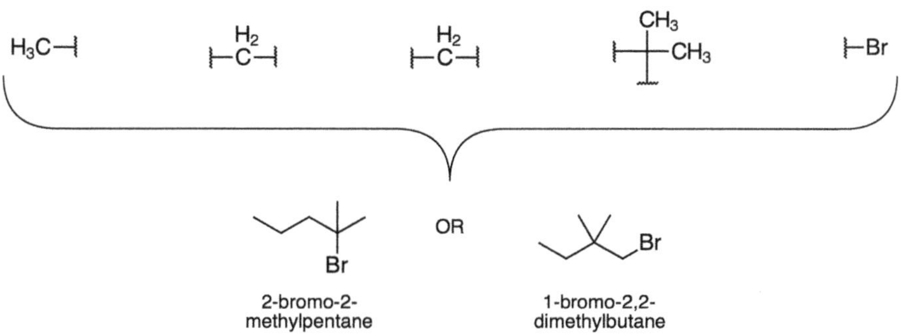

FIGURE 4.20 Fragments deduced from the integration values and the formula, as well as their possible combinations.

We can rule out 1-bromo-2,2-dimethylbutane because it would have two protons of the type H–C–Br, which would appear between 2.5 and 4.5 ppm based on our correlation table (Figure 3.1). No such signal is found in the proton NMR spectrum, so 2-bromo-2-methylpentane must be the correct answer. The fully labeled proton NMR for 2-bromo-2-methylpentane is shown in Figure 4.21.

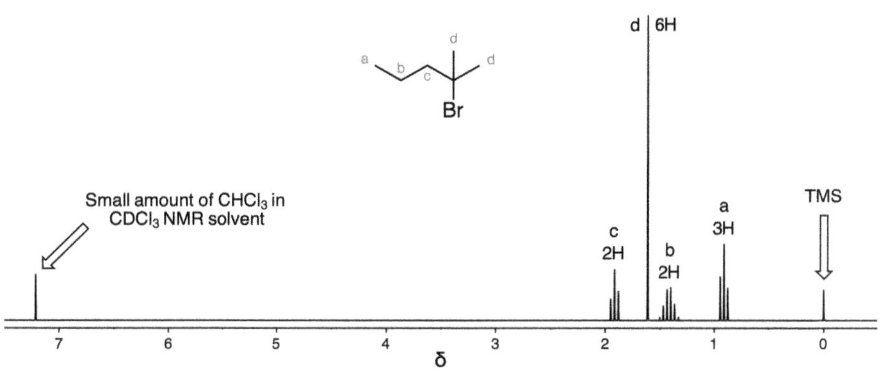

FIGURE 4.21 Labeled proton NMR spectrum of 2-bromo-2-methylpentane.

5 Proton NMR Part IV

Splitting

In the last chapter, we alluded to the fact that proton nuclear magnetic resonance (NMR) signals will often have different shapes (i.e., different numbers of spikes causing the signal). This phenomenon is known as splitting and can also be called multiplicity. It stems from the fact that the spin states of neighboring protons slightly affect the energy of the hydrogen giving rise to the signal.

First, let's define "neighboring" proton. The most common type of neighboring proton is called a vicinal neighbor. "Vicinal" means on adjacent atoms, so a vicinal neighbor is typically a proton on an adjacent carbon (Figure 5.1).

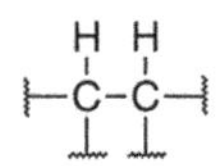

FIGURE 5.1 Vicinal protons.

Let's start to examine some different situations in which protons have varying numbers of vicinal neighbors. The simplest situation is when there are no neighboring hydrogens (Figure 5.2). The hydrogens causing the signal in question are enclosed in boxes. Figure 5.2 shows a methine (CH), a methylene (CH_2), and a methyl (CH_3) group with no vicinal neighbors. Since there are no neighboring hydrogens in each case, there are no effects on the signal's energy, and it appears as a single spike, known as a singlet, in each case. The only difference between the

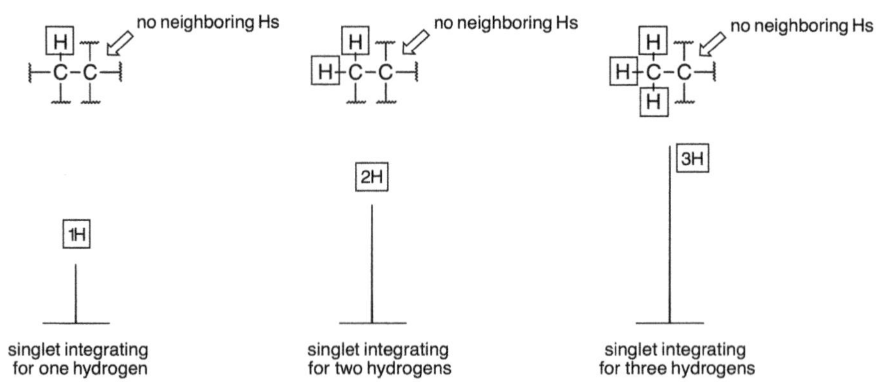

FIGURE 5.2 Groups of hydrogen atoms with no vicinal neighbors.

DOI: 10.1201/9781003783329-5

three scenarios is the integration of each signal (see previous chapter). A methine will integrate for one hydrogen, a methylene for two hydrogens, and a methyl for three hydrogens.

When hydrogen atoms have one vicinal neighbor, that neighbor may be in the spin-up or spin-down state (Figure 5.3). Even in tiny NMR samples, there are enormous numbers of molecules, so some molecules will have this neighboring hydrogen in the spin-up state, while others have it in the spin-down state. These two possible spin states for the neighbor slightly affect the hydrogen(s) causing the signal under consideration, thereby causing it to appear as two spikes (or a doublet) representing those two energy states.

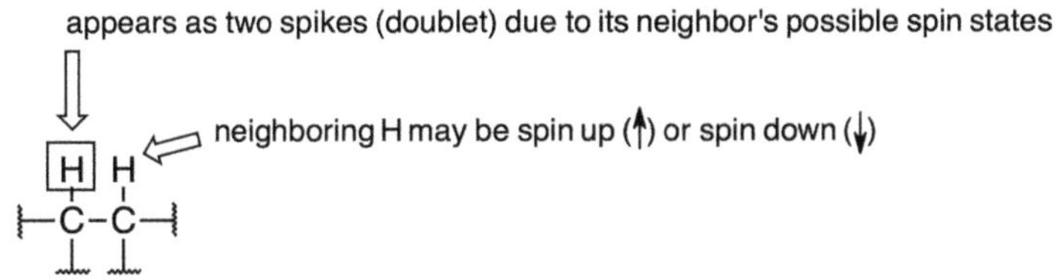

FIGURE 5.3 The effect of a vicinal neighbor on a hydrogen's energy.

Figure 5.4 shows that a methine, methylene, or methyl group with a single neighbor will be split into two spikes, reflecting its two possible energy states: adjacent to a neighbor with spin up and adjacent to a neighbor with spin down. Such a signal is called a doublet, and its two spikes will be of *roughly* equal height because there are equal probabilities of the neighbor being spin up or spin down. Once again, the only difference between the three signals will be in integration.

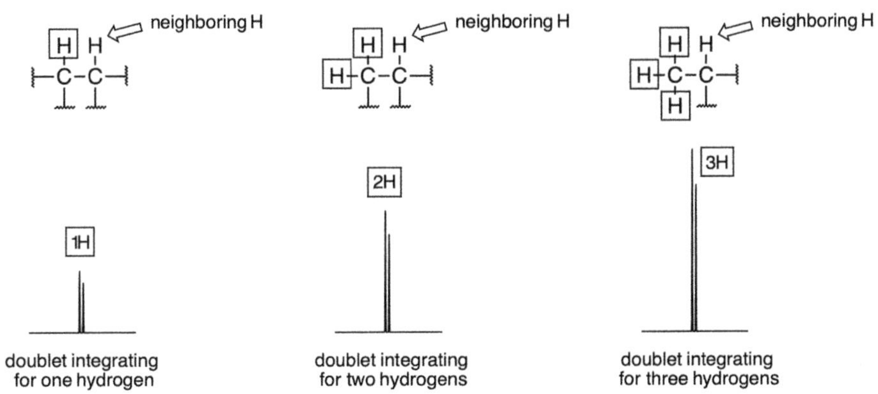

FIGURE 5.4 Groups of hydrogen atoms with one vicinal neighbor.

Splitting is reciprocal, so if a boxed-in hydrogen is split by its vicinal neighbor, then the reverse is also true. However, we methodically consider one type of hydrogen at a time when determining the shape of NMR signals. The neighbors will get their turn to be considered, but in these case studies, we are focusing solely on the hydrogen(s) highlighted in boxes and determining only their splitting patterns.

When hydrogen atoms have two vicinal neighbors, each of which could be spin up or spin down, the situation is a bit more complicated (Figure 5.5). It is possible for both neighbors to be in the spin-up energy state. We could also have only one neighbor spin up, and therefore the other spin down. Lastly, both neighbors could be spin down. The middle state, in which one neighbor is spin up, and one is spin down, can be achieved in two ways. Stated differently, when only one neighbor is spin up, it could be either of the neighbors in that state. Therefore, the hydrogen causing the signal under consideration will be split into three slightly different energy states by its neighbors, resulting in a signal called a triplet. The middle energy state is twice as likely as the others because there are two ways to achieve that state, so the middle spike of the triplet will be roughly twice as tall as the outer spikes.

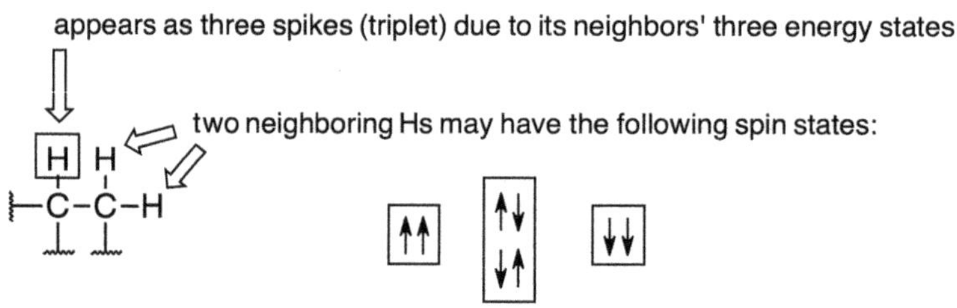

FIGURE 5.5 The effect of two vicinal neighbors on a hydrogen's energy.

Figure 5.6 illustrates methine, methylene, and methyl groups with two neighbors. Each of their signals is split into three spikes, reflecting three unique energy states: adjacent to neighbors that are both spin up, adjacent to one neighbor with spin up and one with spin down, and adjacent to neighbors that are both spin down. These triplet signals consist of three spikes with a height ratio of approximately 1:2:1, reflecting the fact that the middle energy state is twice as probable as the other two states. The only difference between the three signals is their integration value.

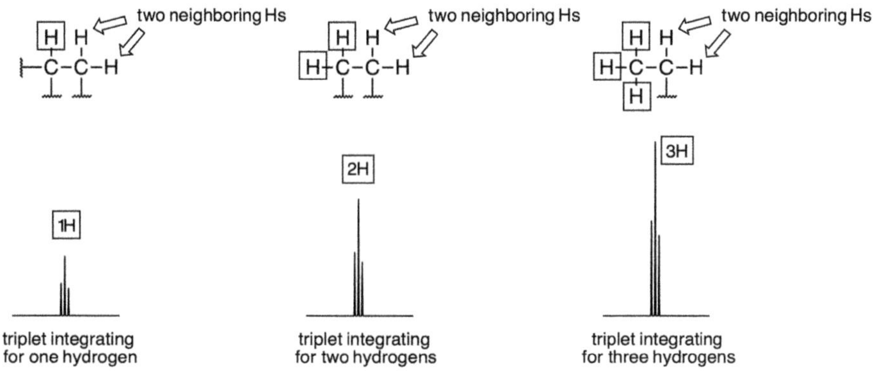

FIGURE 5.6 Groups of hydrogen atoms with two vicinal neighbors.

When hydrogen atoms have three neighbors, the plot thickens. All the neighbors could be spin up, or all the neighbors could be spin down. But there are also two other states in between these extremes. It is possible for only one neighbor to be spin down and for only one neighbor to be spin up. These possibilities are diagrammed in Figure 5.7. Notice that, for the intermediate energy states, each can be achieved in three unique ways. That is to say, if one of the neighbors has a spin different from the others, that oddball could be any of the three.

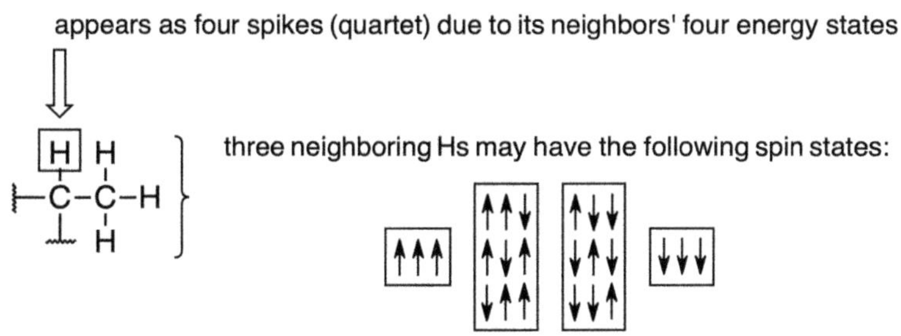

FIGURE 5.7 The effect of three vicinal neighbors on a hydrogen's energy.

When a methine, methylene, or methyl group has three neighbors, its signal is split into four spikes due to the four unique energy states possible, and those four spikes have a height ratio of about 1:3:3:1 because the middle two energy states are three times as probable as the outer two states (Figure 5.8). The signals differ only in their integration values.

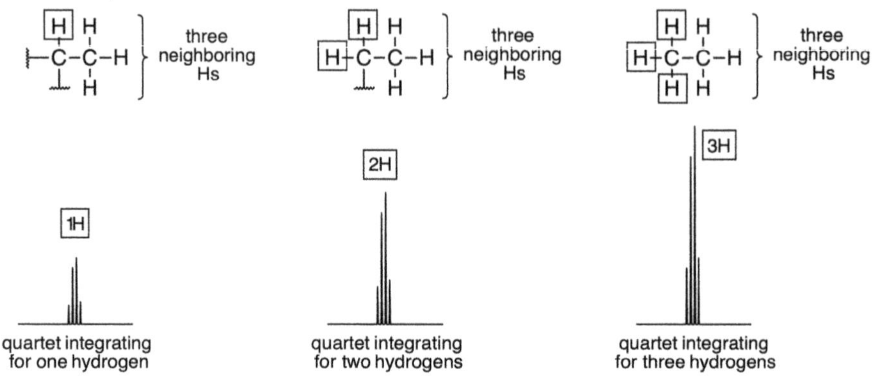

FIGURE 5.8 Groups of hydrogen atoms with three vicinal neighbors.

If we summarize our case studies (Figure 5.9), it becomes apparent that the number of spikes, or multiplicity, of a signal is one more than the number of vicinal neighbors. This gives rise to the "n + 1 rule." In this guideline, n is the number of vicinal neighboring hydrogens. Adding one to that number yields the expected multiplicity for a signal. You can also use this rule in reverse, subtracting one from the multiplicity of a signal to deduce the number of vicinal neighbors.

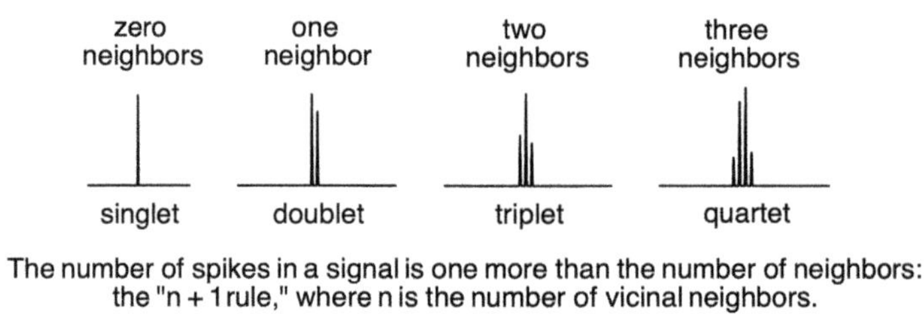

FIGURE 5.9 The n + 1 rule.

There are potential situations beyond what we've covered in our case studies. As long as all of the vicinal neighbors are chemically equivalent, the n + 1 rule will hold. If the multiplicity becomes too great, it may be difficult to accurately count the number of spikes. In such cases, a signal may simply be termed a multiplet, meaning that it has many spikes resulting from many neighbors. If there are vicinal neighbors of more than one type, then splitting patterns become more complex. We'll address this in Chapter 6.

Onc final notc about splitting is that hydrogen-bonding protons (i.e., those on oxygen or nitrogen) typically do not act as true neighbors. Hydrogen-bonding protons can be swapped between molecules, and because they "travel" in this way, they do not usually participate in splitting. Ordinarily, they do not split neighboring hydrogens, and neighboring hydrogens do not split them. However, as we noted in Chapter 3, hydrogen-bonding protons show their presence through broadened signals. It is possible to dilute an NMR sample so much so that hydrogen bonding is negligible because molecules are unlikely to find one another in solution. In such cases, protons on oxygen or nitrogen will then act as "true" neighbors and participate in splitting. However, diluting a sample to this extent is not common practice.

CHAPTER 5 PRACTICE PROBLEMS

1. Let's revisit some of the molecules encountered in the previous chapter (Figures 5.10–5.14). For each type of proton, use the n + 1 rule to predict its signal's multiplicity. All you have to do is specify singlet, doublet, triplet, quartet, or multiplet. However, the answer key will also contain the spectra you've seen previously for these molecules, so that you can actually see those splitting patterns in action.

   a.

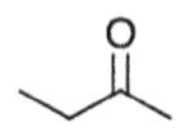

FIGURE 5.10 2-Butanone.

   b.

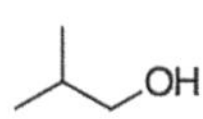

FIGURE 5.11 Isobutyl alcohol.

   c.

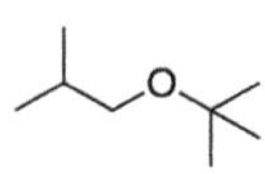

FIGURE 5.12 1-*Tert*-butoxy-2-methylpropane.

   d.

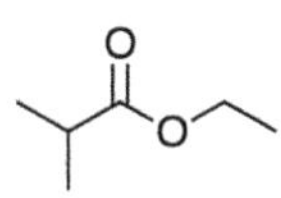

FIGURE 5.13 Ethyl 2-methylpropionate.

e.

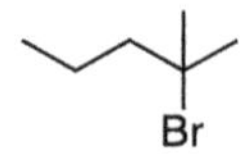

FIGURE 5.14 2-Bromo-2-methylpentane.

2. The structures of propylbenzene and *ortho*-ethyltoluene are given in Figure 5.15, along with their proton NMR spectra. Match each structure with the correct corresponding spectrum based on multiplicity.

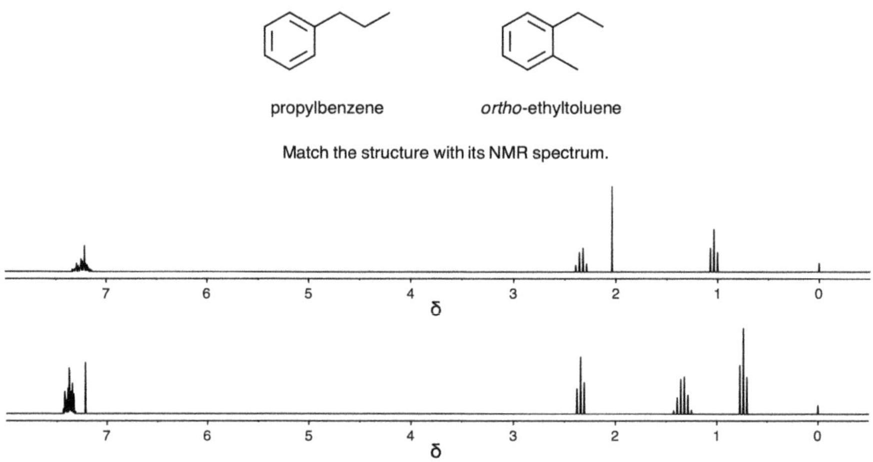

FIGURE 5.15 Two structures to be matched with their corresponding proton NMR spectra.

3. Figure 5.16 contains a proton NMR spectrum. It is the spectrum for either methyl propionate or ethyl acetate. Match the appropriate compound with the NMR spectrum using your knowledge of chemical shift, integration, and splitting.

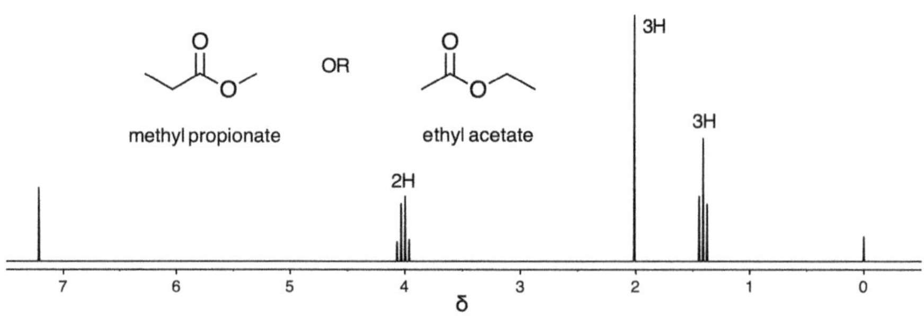

FIGURE 5.16 A proton NMR spectrum to be matched with one of two possible structures.

SOLUTIONS TO CHAPTER 5 PRACTICE PROBLEMS

1.

a. 2-Butanone has three types of protons (Figure 5.17). Let's begin with the methyl group labeled "c." These hydrogens have no vicinal neighbors because there are no protons on the carbonyl carbon. As such, the n + 1 rule tells us to expect a singlet (0 + 1 = 1) for hydrogens "c."

The other methyl group (a) has two vicinal neighbors because it is adjacent to a methylene group (b). The n + 1 rule allows us to predict that the signal for protons "a" should be a triplet (2 + 1 = 3).

In the text of this chapter, we noted that splitting is reciprocal, and we see that here. If the methylene hydrogens (b) are neighbors of the methyl group "a," then it must also be true that the methyl hydrogens (a) are neighbors of methylene "b." Since the methylene hydrogens (b) have three neighbors in the methyl group "a," the n + 1 rule predicts a quartet (3 + 1 = 4) for signal "b."

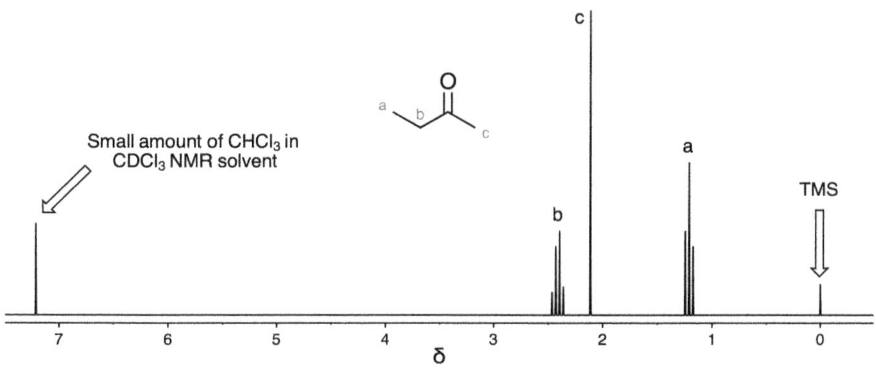

FIGURE 5.17 Proton NMR spectrum of 2-butanone.

b. Isobutyl alcohol contains protons of four different types (Figure 5.18). The protons of the two methyl groups (a) have one vicinal neighbor in the proton of methine (b). Therefore, the n + 1 rule leads us to expect a doublet (1 + 1 = 2) for signal "a."

The methine hydrogen (b) has many neighbors. It has six vicinal neighbors in the methyl groups (a) and two additional vicinal neighbors of a different type (c) in the methylene group. With so many neighbors, especially of different types, we are likely to simply consider the signal for "b" as a multiplet, which implies an unspecified—but large—number of spikes.

The methylene protons (c) have one vicinal neighbor in methine "b." Although there is a second vicinal neighbor in the hydroxyl group's proton (d), that hydrogen-bonding proton is not a true neighbor and does not participate in splitting. Therefore, we expect the methylene signal (c) to appear as a doublet (1 + 1 = 2).

Finally, since the alcohol proton (d) does not participate in splitting, it will be a singlet, but it will also be a *broad* singlet due to its hydrogen bonding.

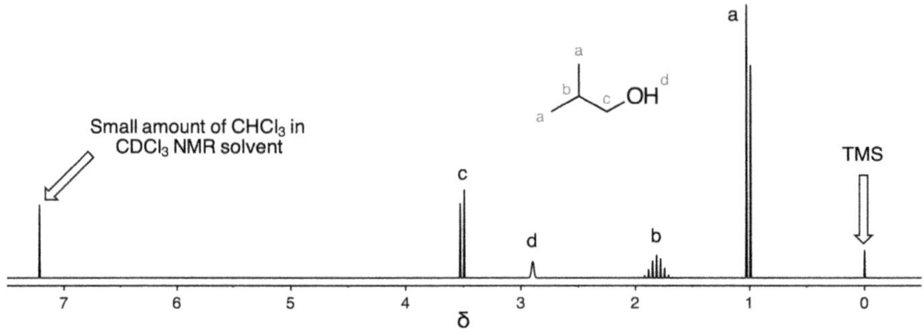

FIGURE 5.18 Proton NMR spectrum of isobutyl alcohol.

c. The compound 1-*tert*-butoxy-2-methylpropane contains four unique types of hydrogens (Figure 5.19). The methyl protons of type “a” have one neighbor in methine “b.” By the n + 1 rule, signal “a” should therefore be a doublet (1 + 1 = 2).

 Methine “b” has many neighbors: six on the two methyl groups of type “a” and two more in the methylene (c). With so many neighbors, especially of different types, it is safest to classify methine “b” simply as a multiplet.

 Methylene “c” has only one neighbor in methine “b,” so the methylene should appear as a doublet (1 + 1 = 2).

 Finally, the three methyl groups of type “d” all reside on a quaternary carbon with no hydrogens, so they have no neighbors. This means that their signal (d) will appear as a singlet (0 + 1 = 1).

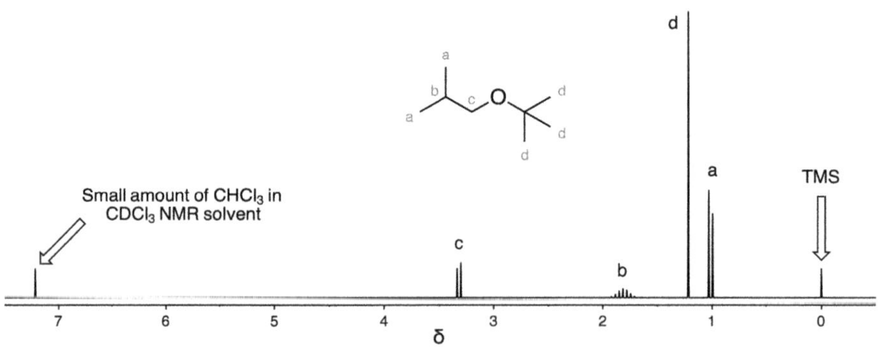

FIGURE 5.19 Proton NMR spectrum of 1-*tert*-butoxy-2-methylpropane.

d. The ester, ethyl 2-methylpropionate, has four distinct types of hydrogens (Figure 5.20). The two equivalent methyl groups of type "a" have one neighbor in the adjacent methine (b). This causes signal "a" to appear as a doublet (1 + 1 = 2).

The methine hydrogen (b) has six equivalent neighbors in the two identical methyl groups "a." Formally, by the n + 1 rule, we can expect a septet (6 + 1 = 7) for signal "b," and if you look closely enough, that is what you will see. However, if you were to simply call this signal a multiplet due to the presence of many neighbors, that would be acceptable, especially because the tiny outermost spikes can be challenging to see.

The methylene (c) has three neighbors in the adjacent methyl group (d), so signal "c" appears as a quartet (3 + 1 = 4). Lastly, since splitting is reciprocal, the methylene hydrogens (c) serve as neighbors to the methyl group "d," causing its signal to be split into a triplet (2 + 1 = 3).

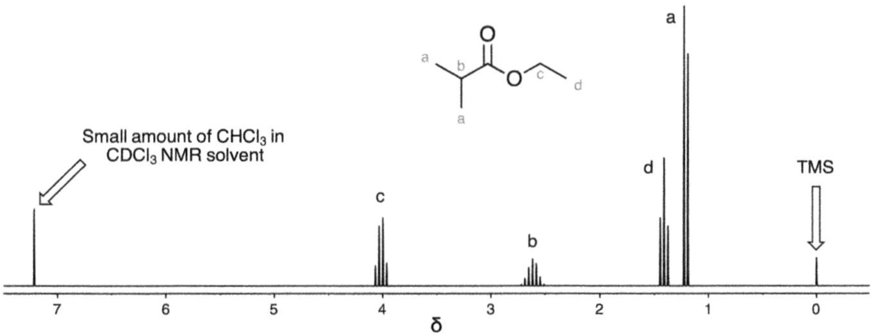

FIGURE 5.20 Proton NMR spectrum of ethyl 2-methylpropionate.

e. 2-Bromo-2-methylpentane has four unique types of hydrogens as well. Methyl group "a" has two neighbors in the adjacent methylene (b, Figure 5.21). Therefore, the signal for protons "a" is a triplet (2 + 1 = 3).

The methylene "b" has many neighbors, and they are of different types. There are three neighbors in the adjacent methyl group (a) and two different neighbors in the adjacent methylene (c). Consequently, we may refer to signal "b" as a multiplet.

The other methylene group (c) has only two neighbors (b). Therefore, the signal for protons "c" will be a triplet (2 + 1 = 3). Finally, the two equivalent methyls of type "d" have no neighbors because they stem from a carbon without hydrogens. As a result, signal "d" is a singlet (0 + 1 = 1).

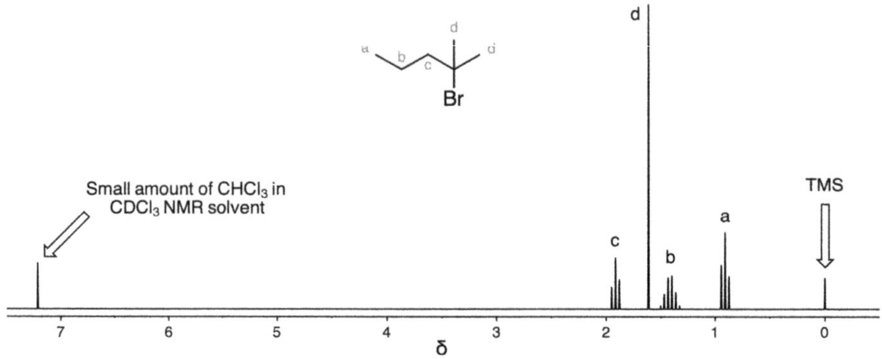

FIGURE 5.21 Proton NMR spectrum of 2-bromo-2-methylpentane.

2. In the spectra of both compounds, we see complex, overlapping signals in the aromatic region (6.5–8 ppm). We know this because the signals are not symmetrical as we see for singlets, doublets, triplets, quartets, etc. Note that the residual solvent signal (the tiny amount of $CHCl_3$ present in the $CDCl_3$ solvent) appears as a singlet at about 7.24 ppm. In one spectrum, it is clear. In the other, it overlaps with the aromatic proton signals. In any event, we will not be able to gain much insight by comparing the aromatic signals for these substances.

Instead, to differentiate the spectra, we must focus on the substituents connected to the aromatic rings. In the case of propyl benzene, the propyl group gives rise to three signals (Figure 5.22). Its methyl group (c) and one of its methylenes (a) appear as triplets because both have two neighbors located at methylene "b." The methylene "b" has many neighbors (the two at "a" plus the three at "c"), so it appears as a multiplet.

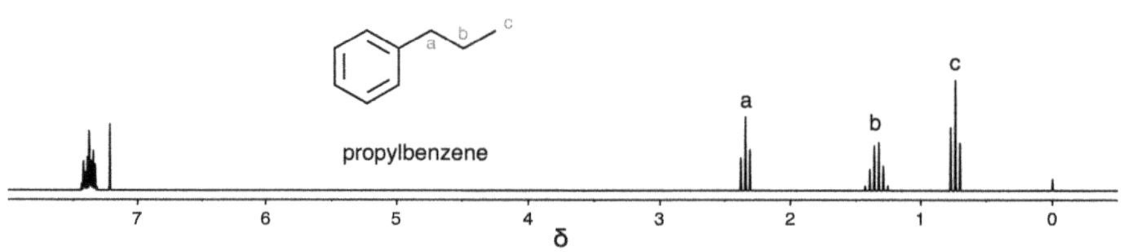

FIGURE 5.22 Proton NMR spectrum of propylbenzene.

In the case of *ortho*-ethyltoluene, one methyl group (c) has no neighboring hydrogens, so it appears as a singlet (Figure 5.23). The other methyl group (b) has two neighbors at methylene “a,” so the signal for “b” is a triplet. The methylene (a) has three neighbors in the methyl group “b,” so it is a quartet.

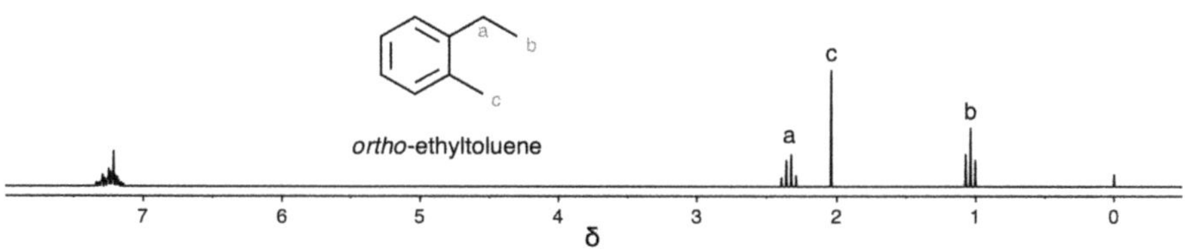

FIGURE 5.23 Proton NMR spectrum of *ortho*-ethyltoluene.

3. It is best to begin this problem by predicting each signal’s chemical shift, integration, and splitting (Figure 5.24). The chemical shifts are derived from the correlation table in Figure 3.1. Integrations simply refer to the number of hydrogen atoms causing each signal. The splitting of each signal is determined by counting its vicinal neighbors and using the n + 1 rule.

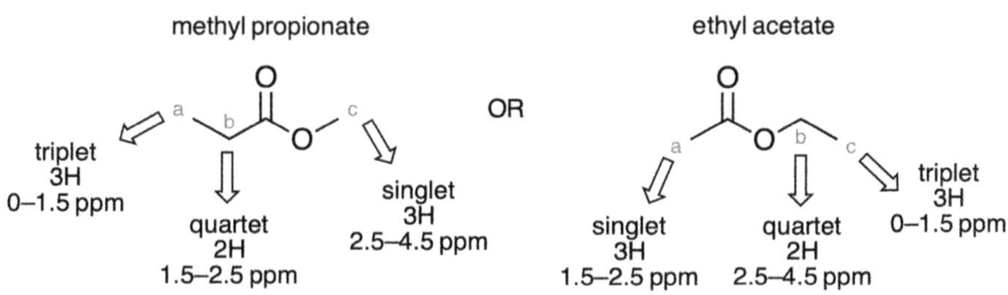

FIGURE 5.24 The expected proton NMR signals for methyl propionate and ethyl acetate.

There are enough similarities between the expected signals that we have to be strategic in how we compare them. Both molecules will produce a triplet integrating for three hydrogens, a quartet integrating for two hydrogens, and a singlet integrating for three hydrogens. Therefore, we cannot differentiate them based on these features. The clearest distinction between the two structures is in which of the signals is the most deshielded (i.e., has the highest chemical-shift value). In methyl propionate, it is the singlet for three hydrogens (c) that is most deshielded; however, in ethyl acetate, it is the quartet for two hydrogens (b) that is most deshielded. On the basis of this difference, we can assign the spectrum to ethyl acetate (Figure 5.25). Ethyl acetate is the correct answer to this question.

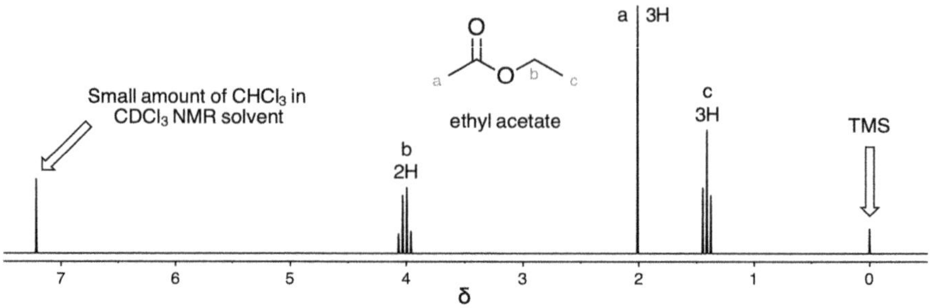

FIGURE 5.25 Proton NMR spectrum of ethyl acetate.

Although methyl propionate is the *incorrect* answer to this question, it can be useful to see what its proton NMR spectrum would look like (Figure 5.26). Notice that it would have the singlet for the methyl group (c) at the most deshielded (i.e., highest chemical-shift value) location.

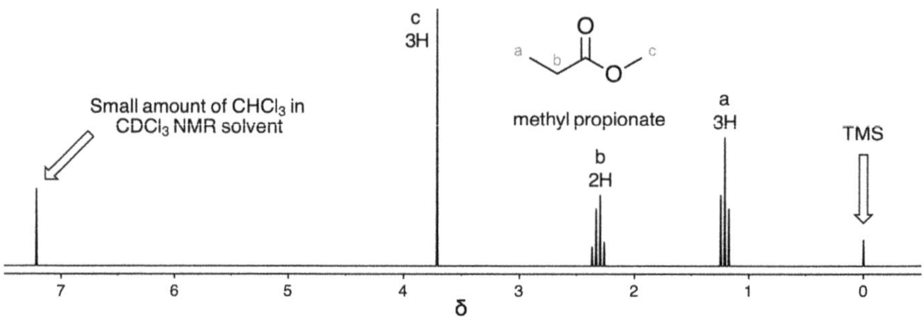

FIGURE 5.26 Proton NMR spectrum of methyl propionate.

6 Proton NMR Part V
Advanced Topics

In order to understand some unusual splitting patterns that we can encounter in proton nuclear magnetic resonance (NMR), we need to understand a bit more about the coupling between protons that leads to splitting. In the previous chapter, we saw that, when a proton has a single neighbor, this neighbor may be spin up or spin down, so the signal is split into a doublet. This can be represented by a splitting tree in which we show the proton's signal being divided into two spikes of roughly equal height (Figure 6.1). The distance between the spikes is known as the coupling constant, or *J* value, which is typically measured in Hertz (Hz).

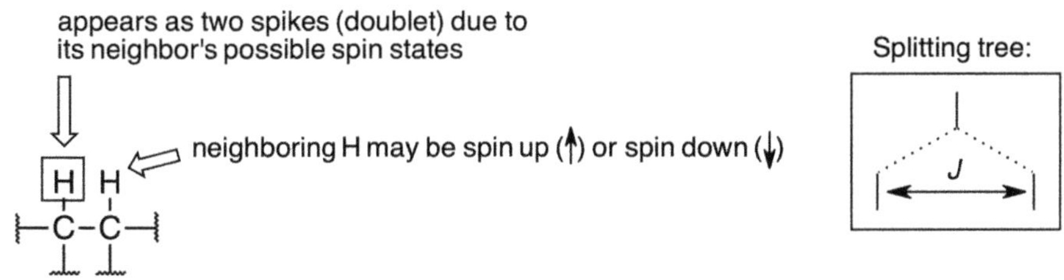

FIGURE 6.1 A splitting tree for a proton or protons with a single neighbor.

Each neighbor present can be in the spin-up or spin-down energy state, so each neighbor will exert this same effect—splitting a proton's possible energy states in two. When a proton has two chemically equivalent neighbors, the first splits the signal into two spikes, and the second splits each of those spikes into two (Figure 6.2). If the two neighbors are, in fact, chemically equivalent, then the *J*

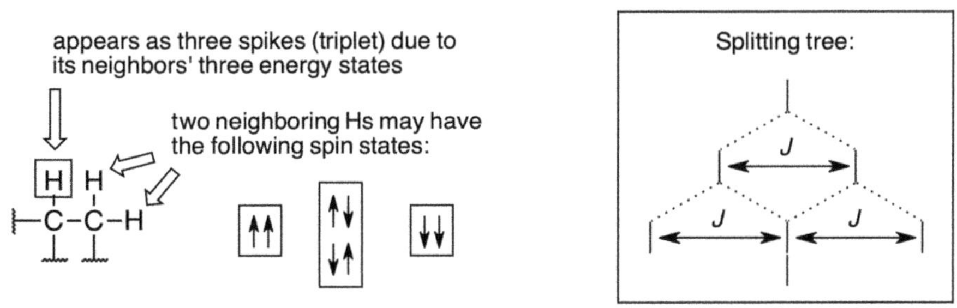

FIGURE 6.2 A splitting tree for a proton or protons with two neighbors.

DOI: 10.1201/9781003783329-6

values will be identical, meaning that the central branches of the splitting tree meet. When this happens, the heights are additive. This results in the triplet pattern we saw in the previous chapter (Figure 5.5). The fact that the middle spike is about twice as tall as the outer spikes reflects the fact that the middle energy state is twice as probable.

When a third chemically equivalent neighbor is present, it splits each of the spikes of the triplet into two (Figure 6.3). Again, if it is, in fact, indistinguishable from the other neighbors, it will couple with an identical *J* value so that the central branches of the splitting tree merge. When they do, one height unit adds to two height units, giving central spikes roughly three times as tall as the outer ones. Recall that this is reflective of those central energy states being three times as likely (Figure 5.7).

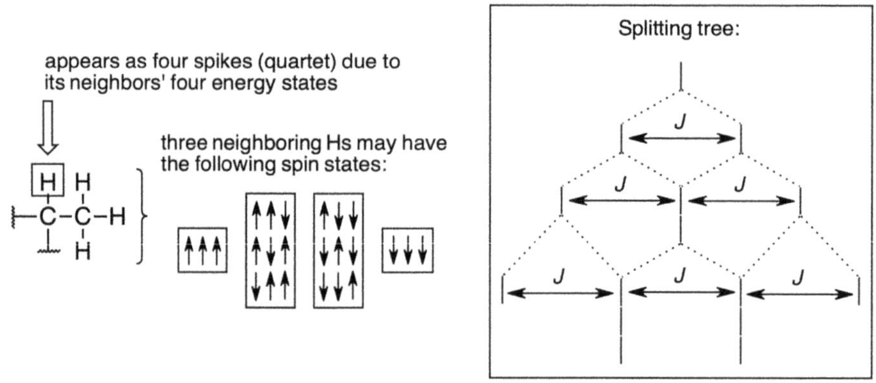

FIGURE 6.3 A splitting tree for a proton or protons with three neighbors.

These splitting trees have explained the multiplicities we saw in the previous chapter in a different way. That's perhaps helpful, but we don't see the true value of splitting trees until we consider situations in which there are chemically *inequivalent* neighbors. As we'll see momentarily, these scenarios will *not* follow the n + 1 rule. For example, let's consider what happens if a proton has two chemically *inequivalent* neighbors. The first splits the signal into a doublet with a certain *J* value. The second will split each of those spikes into two, but since this neighbor is different from the first, it will couple with a different *J* value. Consequently, it is quite possible that the central branches of the splitting tree will not merge in this case, which would result in four spikes of roughly equal height (Figure 6.4). We have to be careful how we describe this. It is *not* a quartet. Although a quartet does have four spikes, it also has a distinctive height ratio of 1:3:3:1 for those four spikes. This signal has a height ratio of 1:1:1:1, which makes it something different. It is really a pair of doublets, so we call it a doublet of doublets.

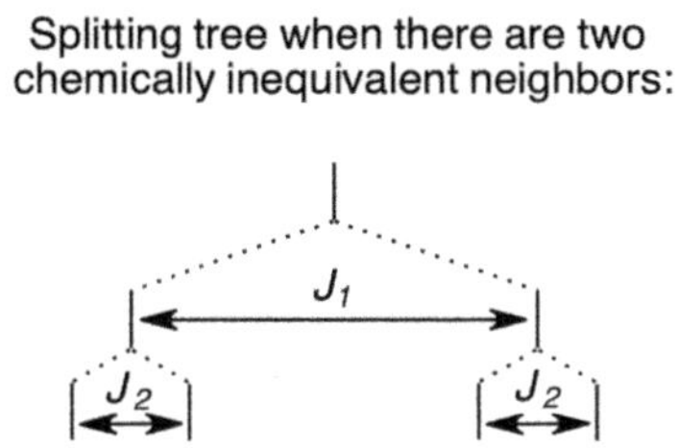

FIGURE 6.4 A splitting tree for a proton or protons with two chemically inequivalent neighbors.

Note that it doesn't matter in what order you consider the two couplings. However, if you draw the one with the smaller J value first, the branches of the splitting tree will then cross, which is less aesthetically pleasing and less clear (Figure 6.5). For this reason, we usually consider the couplings in order of decreasing J value, simply because it results in clearer diagrams.

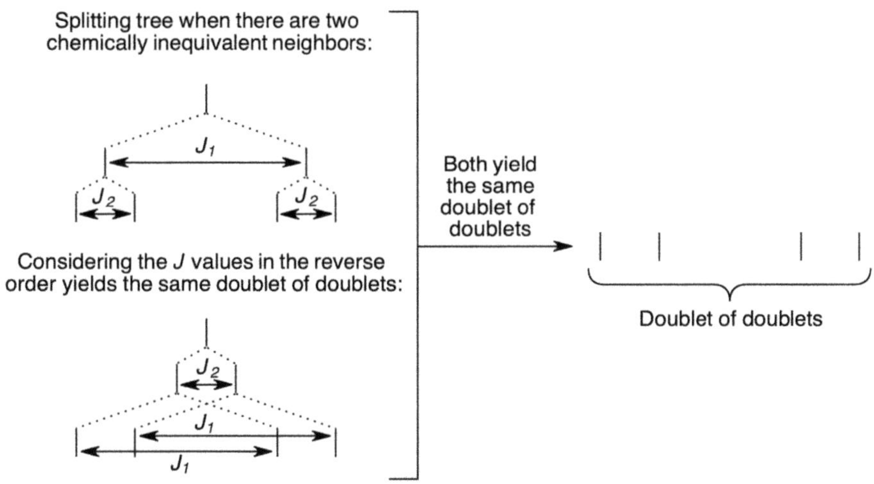

FIGURE 6.5 A splitting tree for a proton or protons with two chemically inequivalent neighbors showing the couplings in the reverse order.

How could a proton or protons end up with two chemically inequivalent neighbors, though? There are a few possibilities. It is conceivable that a proton or protons could have a methine on each side and that these methines could be different because of asymmetry within the molecule (Figure 6.6).

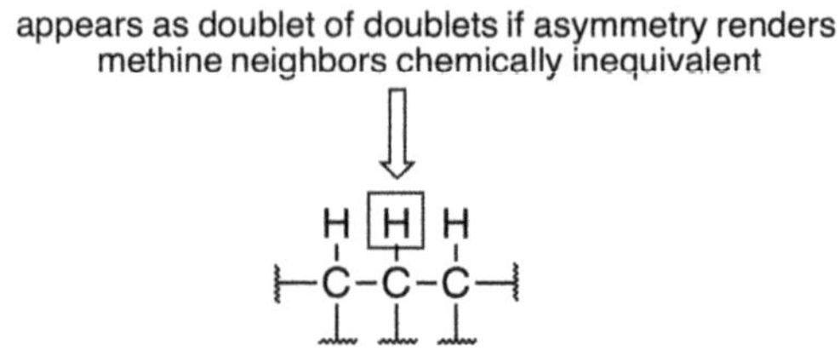

FIGURE 6.6 A proton or protons with two different methine neighbors.

It is also possible that long-range splitting could play a role. Thus far, we have only considered vicinal neighbors, or protons on adjacent carbons, but in some situations where there is conformational constraint, longer-range splitting is possible. For example, on an aromatic ring, there can be splitting between *ortho*, *meta*, and *para* "neighbors." *Ortho* splitting has the largest coupling constant (typically 6–9 Hz). *Meta* and *para* splitting have smaller coupling of constants, typically in the range of 1–3 and 0–1 Hz, respectively. So, an aryl proton with, for example, both *ortho* and *meta* neighbors would also result in a doublet of doublets (Figure 6.7).

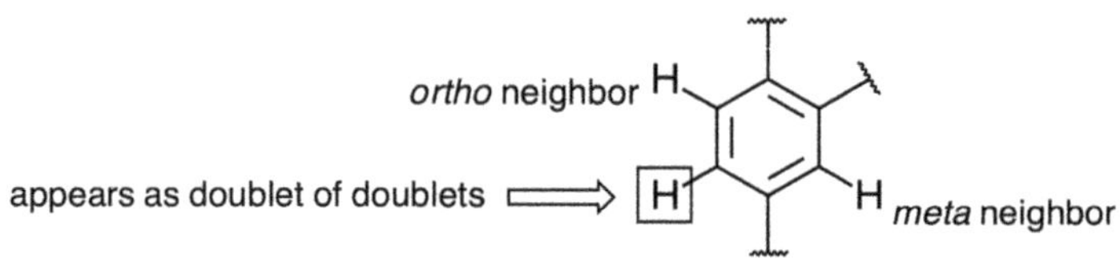

FIGURE 6.7 An aryl proton with *ortho* and *meta* neighbors.

Another scenario that could result in a doublet of doublets is diastereotopic methylene hydrogens with a methine neighbor. Let's unpack that situation. To this point, we've considered protons stemming from the same carbon to be chemically equivalent. This is often, but not always, the case. When there is at least one chiral center in the molecule, methylene hydrogens near the chirality will no longer be chemically equivalent.

Consider the chiral molecule (*R*)-4,4-dimethyl-2-pentanol shown in Figure 6.8. One methylene hydrogen is closer to the hydroxyl group than the other. While it is true that we could look at different conformations of this molecule, in any conformation we choose to examine, the methylene hydrogens will be at different distances from the groups near them. As a result, they are in different chemical environments.

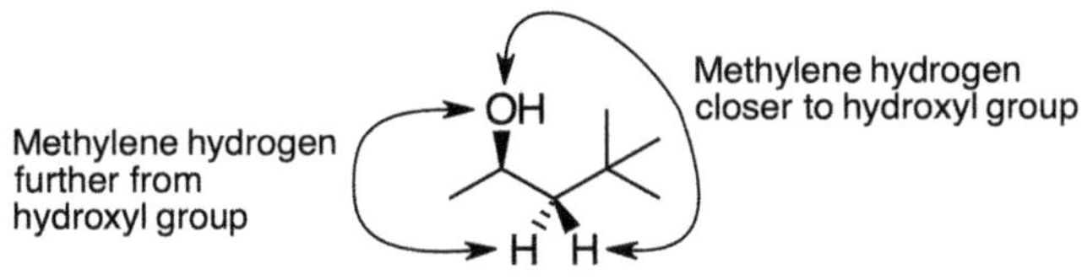

FIGURE 6.8 (*R*)-4,4-Dimethyl-2-pentanol.

Such methylene hydrogens are called diastereotopic, and they cause different signals since they are chemically inequivalent. Furthermore, since they are distinct types of hydrogens, they will act as geminal neighbors to one another, where geminal means stemming from the same atom (Figure 6.9). Geminal coupling (J = 12–15 Hz) is typically greater than the vicinal coupling (J = 6–8 Hz) caused by the adjacent methine hydrogen.

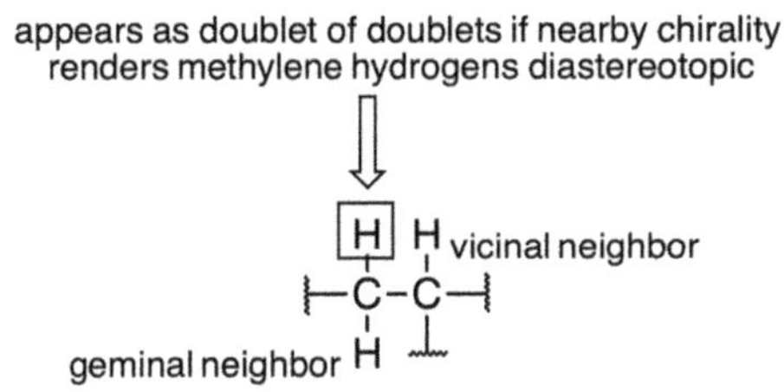

FIGURE 6.9 Diastereotopic methylene hydrogens with a methine neighbor.

A final note involves the potential deformation of signals from their ideal or "textbook" shape. When two coupled signals are close enough in chemical shift, a phenomenon known as second-order coupling, or roofing, occurs. The spikes of each signal that are closest to one another grow in intensity, while the outermost spikes of the pair diminish in size, making the two signals together take on the appearance of a peaked roof. Consider the proton NMR spectrum of (*R*)-4,4-dimethyl-2-pentanol (Figure 6.10). The inset for the diastereotopic methylene hydrogens illustrates this property because the interior peaks are slightly more pronounced than the exterior ones. The effect becomes more intense the closer the two signals are in chemical shift.

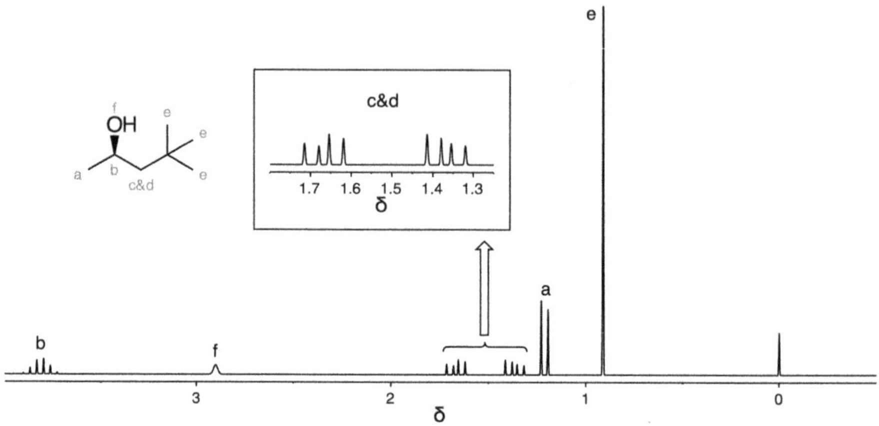

FIGURE 6.10 Proton NMR spectrum of (*R*)-4,4-dimethyl-2-pentanol.

CHAPTER 6 PRACTICE PROBLEMS

1. The structure of catechol is shown in Figure 6.11.

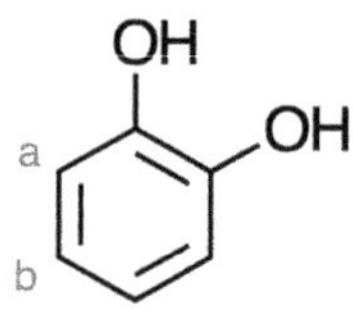

FIGURE 6.11 Catechol.

   a. Draw a splitting tree for the proton at location "a" on the ring.
   b. Draw a splitting tree for the proton at location "b" on the ring.
   c. Predict the proton NMR spectrum's appearance.
   d. What might happen if the signals for the protons on the aromatic ring are close in chemical shift?

2. Assign letters to the different types of protons in [(*R*)-2-chloropropyl]benzene (Figure 6.12). Then, label the peaks in the proton NMR spectrum with the corresponding letters.

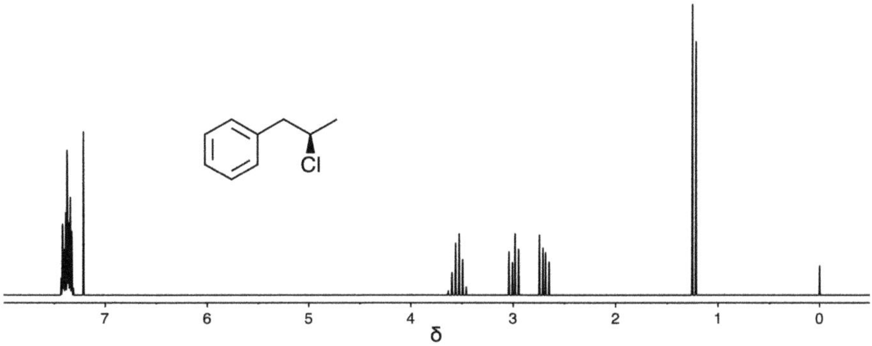

FIGURE 6.12 Proton NMR spectrum of [(*R*)-2-chloropropyl]benzene.

3. Predict the proton NMR spectrum for 3-buten-2-amine (Figure 6.13). For this question, you'll also need to know that: (a) the coupling constant for geminal neighbors on an sp^2 carbon is 0.5–3 Hz, (b) the coupling constant for *cis* vicinal neighbors on an alkene is 8–12 Hz, and (c) the coupling constant for *trans* vicinal neighbors on an alkene is 14–17 Hz.

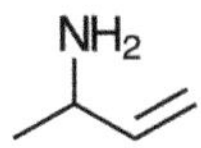

FIGURE 6.13 3-Buten-2-amine.

SOLUTIONS TO CHAPTER 6 PRACTICE PROBLEMS

1.
   a. This molecule has an internal plane of symmetry, so there are only three types of protons: those labeled "a," "b," and "c" (Figure 6.14). The proton at site "a" will participate in two couplings. It will couple with its *ortho* neighbor (b) and with its *meta* neighbor [b', which is chemically equivalent to "b" but is given a prime (') to differentiate it

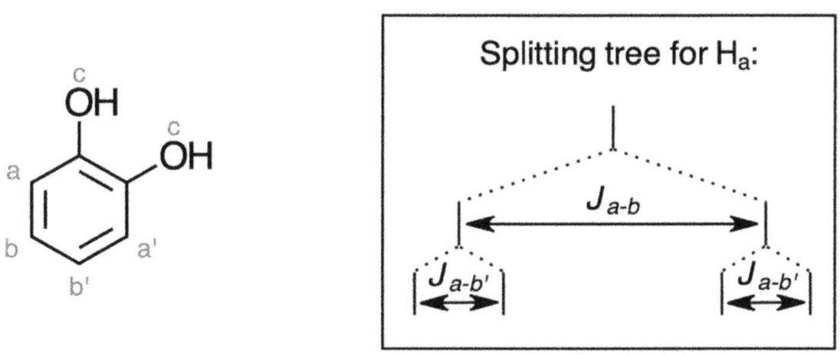

FIGURE 6.14 A splitting tree for proton "a."

in this regard]. The *ortho* splitting ($J_{a\text{-}b}$) will be larger (about 6–9 Hz). The *meta* splitting ($J_{a\text{-}b'}$) will be smaller (about 1–3 Hz). The splitting tree will be cleaner if we consider the larger *ortho* splitting first. Then, each of the resulting spikes is split into two by the *meta* coupling. The resulting signal will be a doublet of doublets.

Since the two protons labeled "a" are chemically equivalent, they will cause one signal, and it will be a doublet of doublets. Notice, though, that to determine the signal's splitting pattern, we consider just one proton labeled "a." The other proton of this type (a') is in the exact same situation, being split by its *ortho* neighbor (b') and its *meta* neighbor (b). Also, keep in mind that chemically equivalent protons will not split one another, so the two protons of type "a" do not couple.

b. Proton "b" similarly engages in two coupling events. It will couple with its *ortho* neighbor (a) and its *meta* neighbor (a'). Since both protons of type "b" are chemically equivalent, they do not split one another. Again, we'll get a clearer splitting tree if we consider the coupling events in order of decreasing coupling constant (Figure 6.15). The *ortho* J value (6–9 Hz) is larger than the *meta* J value (1–3 Hz).

Since the two protons labeled "b" are chemically equivalent, they cause one signal, and it is a doublet of doublets. Notice, though, that to determine the signal's splitting pattern, we consider just one proton labeled "b." The other proton of this type (b') is in the exact same situation, being split by its *ortho* neighbor (a') and its *meta* neighbor (a).

c. Using what we learned in parts (a and b) above, we can predict that protons "a" will yield a doublet of doublets in the aromatic region (6.5–8 ppm), and the protons labeled "b" will also appear as a doublet of doublets in the same region. The alcohol protons (c) will be a broad singlet with a highly variable chemical shift that depends on the concentration of the sample. Each signal is due to two hydrogen atoms, so all three signals will have the same integration value.
d. We know from part (c) above that we expect two doublets of doublets, each having four spikes of roughly equal height, in the aromatic region. If the chemical shift difference between the two signals is not large, then we will see second-order coupling, also known as roofing. This will cause the innermost spikes of the pair of signals to grow in intensity, while the outermost spikes of the pair shrink.

In fact, that is what we see in Figure 6.16. In the inset enlargement of the aromatic region, there are eight spikes as we expect for two doublets of doublets. However, they are not all of equal height. We are seeing roofing.

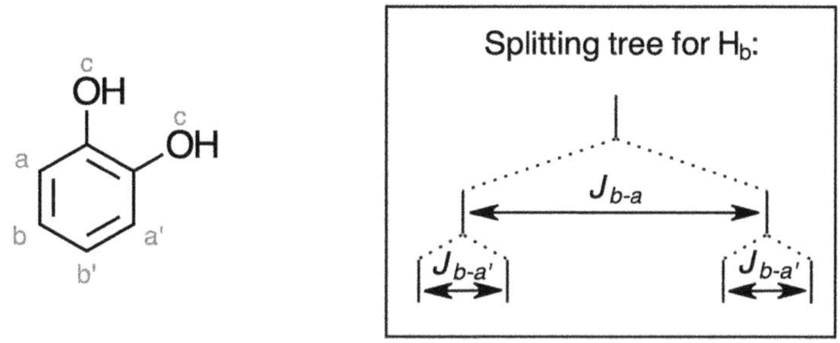

FIGURE 6.15 A splitting tree for proton "b."

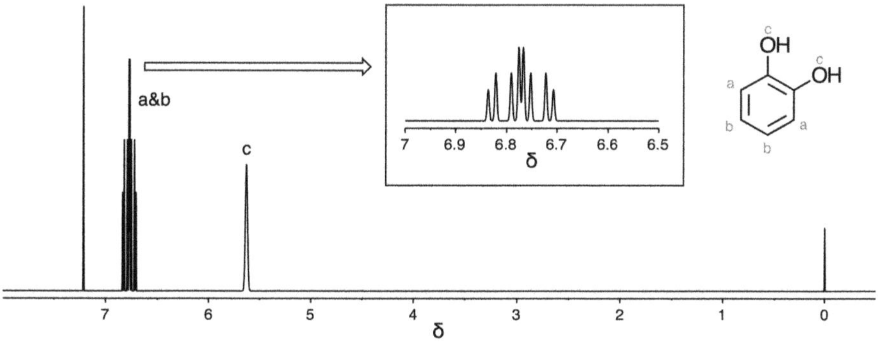

FIGURE 6.16 Proton NMR spectrum of catechol.

Note that the signals at 0 and 7.24 ppm are merely TMS and residual chloroform in the deuterated chloroform NMR solvent, respectively.

2. [(*R*)-2-Chloropropyl]benzene contains several types of protons (a–g in Figure 6.17). The aromatic ring alone has three types of protons (a–c). However, as is often the case, especially with monosubstituted aromatic rings, these three signals overlap to form a messy multiplet. Therefore, any more specific assignment of the aryl protons is not easily made. Remember that the adjacent singlet at 7.24 ppm is merely due to residual $CHCl_3$ in the $CDCl_3$ NMR solvent.

As we continue to move along the x-axis from high to low chemical shift, we next encounter the signal for proton "f." After the aryl protons, this is the next most deshielded proton because it is on the carbon bonded to chlorine. Proton "f" has many neighbors. Those neighbors are of multiple types, so it will be split into a complex multiplet.

The methylene protons (d and e) are somewhat deshielded since they are adjacent to an aromatic ring and are also adjacent to the carbon bearing a halogen. Additionally, they are diastereotopic since there is a chiral center in the molecule,

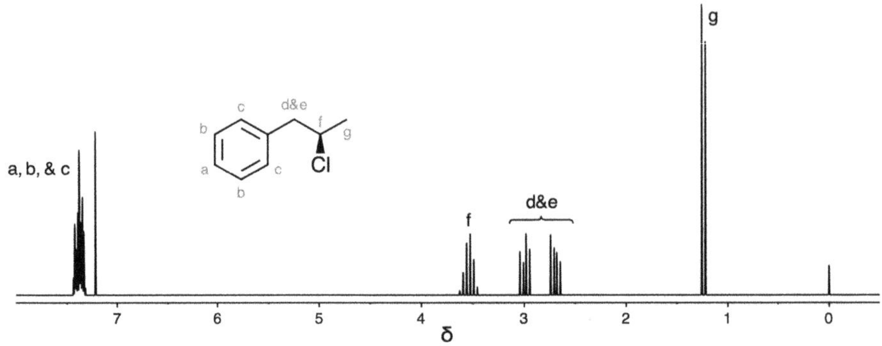

FIGURE 6.17 Labeled proton NMR spectrum of [(*R*)-2-chloropropyl]benzene.

and they are certainly close enough to it to feel its effect. As a result, each of the methylene protons will yield its own signal. Furthermore, each of these signals will be a doublet of doublets because the proton will be split by its geminal neighbor and by its vicinal neighbor (f).

Finally, the methyl protons (g) are furthest from the functionality and are therefore the most shielded. They are split into a doublet by their sole neighbor (f).

Remember that the signal at 0 ppm is merely the TMS internal standard.

3. 3-Buten-2-amine contains six types of protons (a–f, Figure 6.18). The trickiest part is identifying the two terminal vinyl hydrogens (d and e) as different. One (e) is *cis* to the substituent that begins with carbon "b," while the other (d) is *trans* to that substituent. Since there is no free rotation about a carbon-carbon double bond, proton "e" will always be *cis* to the substituent, while proton "d" will always be *trans* to the substituent. This renders "d" and "e" chemically inequivalent, so they will give rise to two separate signals.

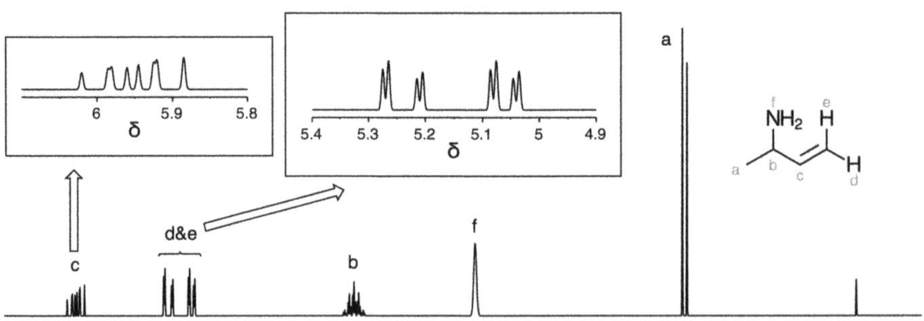

FIGURE 6.18 Proton NMR spectrum of 3-buten-2-amine.

Now that we have identified the number of signals to expect, we can begin to place them on the chemical shift axis and consider their splitting patterns. Protons "c," "d," and "e" will all fall within the vinyl region (4.5–6.5 ppm). Beyond that, it would be challenging to predict their exact ordering on the x-axis. Protons "d" and "e" will each be doublets of doublets, which is highlighted in one inset in Figure 6.18. Each one is split into a doublet by its vicinal neighbor (c). Then each of those two spikes is further split into two by the geminal neighbor. The geminal coupling is smaller than the vicinal coupling.

Proton "c" will have an even more complex splitting pattern. Its signal will be split into two spikes by its *trans* vicinal neighbor (e, J = 14–17 Hz). Each of those spikes will be split into two by its *cis* vicinal neighbor (d, J = 8–12 Hz). Finally, each of those four spikes of roughly equivalent height will then be split into two by the regular vicinal neighbor (b, J = 6–8 Hz). That yields a total of eight spikes, a pattern known as a doublet of doublet of doublets. In the inset in Figure 6.18, you can actually see all eight spikes.

Proton "b" is on a carbon bonded to nitrogen, so it should appear between 2.5 and 4.5 ppm. It has many neighbors of multiple types, so it will simply be a complex multiplet.

The methyl group (a) is the most shielded signal since it is furthest from functionality, and it is split into a doublet by its sole neighbor (b).

The amine protons (f) are capable of hydrogen bonding, so their signal is broadened and has a highly variable chemical shift that depends on concentration.

Signals b–e all integrate for one hydrogen apiece, while the amine protons (f) integrate for two, and the methyl hydrogens (a) integrate for three.

7 Carbon-13 NMR

While the most abundant isotope of carbon—carbon-12—is not nuclear magnetic resonance (NMR) active, carbon-13 (^{13}C) *is* NMR active. Carbon-13 accounts for only about 1.1% of all carbon, so unlike proton NMR, in which the element's most abundant isotope is yielding NMR information, in carbon-13 NMR, only a relatively rare isotope is providing information. As a result, carbon-13 NMR spectra take a bit longer to acquire, especially with the small sample sizes that might be used in research settings where material is precious.

There are similarities in how proton and carbon-13 NMR spectra are analyzed, but there are also important differences in their acquisition that will make the interpretation of ^{13}C NMR much simpler.

The first piece of information we will consider is the number of signals to expect in the carbon-13 NMR spectrum. This is directly analogous to our experience with proton NMR. Each unique type of carbon atom will yield its own signal in the ^{13}C spectrum. Carbons that are chemically equivalent will yield a single, combined signal.

The next important piece of information is the chemical shift. As with proton NMR, the location of a signal on the x-axis tells us something about the chemical environment in which the nucleus finds itself. However, the chemical shift range is much wider for carbon-13 NMR, spanning 0 ppm to over 200 ppm (Figure 7.1). Consequently, the accidental overlap of distinct signals, which we sometimes see in proton NMR, is much less likely in carbon-13 NMR. The internal standard

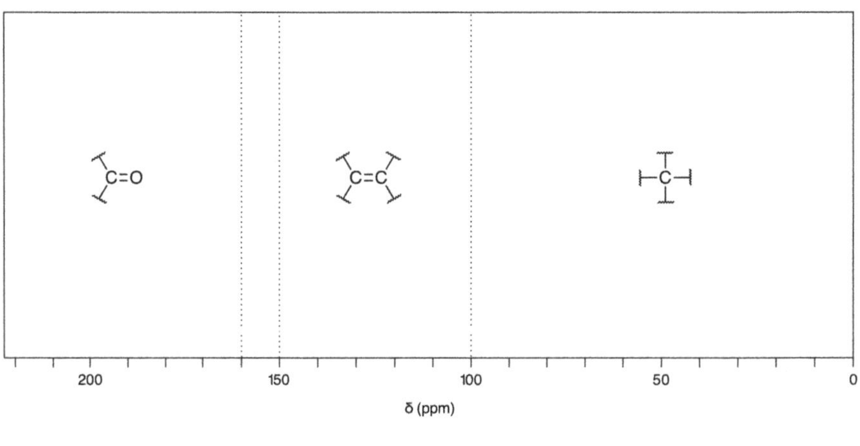

FIGURE 7.1 Correlation table for carbon-13 NMR.

DOI: 10.1201/9781003783329-7

[TMS or tetramethylsilane, $(CH_3)_4Si$] produces a single carbon signal that once again marks 0 ppm.

In general, sp^3 hybridized carbons tend to appear below 100 ppm. Carbons participating in carbon-carbon double bonds (i.e., alkene and aromatic carbons) usually fall between 100 and 150 ppm, while carbonyl carbons typically appear above 160 ppm.

Integration and splitting are not terribly important considerations in carbon-13 NMR. Since protons can have two spin states (spin up or spin down), if we took no measures to prevent it, the signals for carbon atoms would be split not only by the attached protons but also by those on neighboring carbons. This would result in complex splitting patterns and overlapping signals (because protons split the carbon signals with a large coupling constant, making the distance between spikes large). Since this is not desirable, the most typical carbon-13 NMR spectrum is a *proton-decoupled* carbon-13 spectrum. The term "proton-decoupled" means that the protons are constantly irradiated through the data acquisition in order to keep them in the excited state. Since they are therefore always in only one energy state (the excited state) instead of the normal two possible energy states, they no longer split any of the carbon signals. For this reason, all the carbon-13 signals of the analyte appear as singlets, and we have *no* splitting with which to concern ourselves.

The process of proton decoupling renders integration values unreliable, so ^{13}C spectra do *not* typically have integration values either. However, we will sometimes be able to use integration loosely if we notice that a signal is substantially larger than the others.

A quick note about the solvent is warranted. The carbon atom in deuterated chloroform ($CDCl_3$) will appear in the carbon-13 NMR spectrum. Its signal will be split by the attached deuterium. Deuterium actually has three spin states (known as −1, 0, and 1), so it will split the carbon signal into three spikes of equivalent height because there is an equal probability of the deuterium atom being in any of those three states. This signal appears at about 77 ppm in the ^{13}C spectrum. Despite having three spikes, it is not a true triplet because it lacks the 1:2:1 height ratio of a triplet.

Let's practice analyzing a carbon-13 spectrum together. Consider the molecule propiophenone (Figure 7.2). This structure has carbon atoms that will give signals in each of the three chemical shift ranges. Before we get into a detailed analysis, we should remember that the signal at 0 ppm is TMS (the internal standard), and the signal at 77 ppm is the deuterated chloroform NMR solvent. These two signals are unrelated to our analyte, so we can effectively ignore them.

Carbons "f" and "g" are sp^3 hybridized and therefore produce signals less than 100 ppm. Atoms closer to a functional group will typically be more deshielded, so we can surmise that carbon "f" corresponds to the more deshielded signal near 35 ppm since it is adjacent to the ketone's carbonyl. Carbon "g" is further from the functional group and therefore correlates with the more shielded signal below 10 ppm.

Carbon "e" is a carbonyl carbon and therefore appears in the range of 160–220 ppm. The carbons of the aromatic ring are expected between 100 and 150 ppm. Although integration is not strictly reliable in proton-decoupled carbon-13

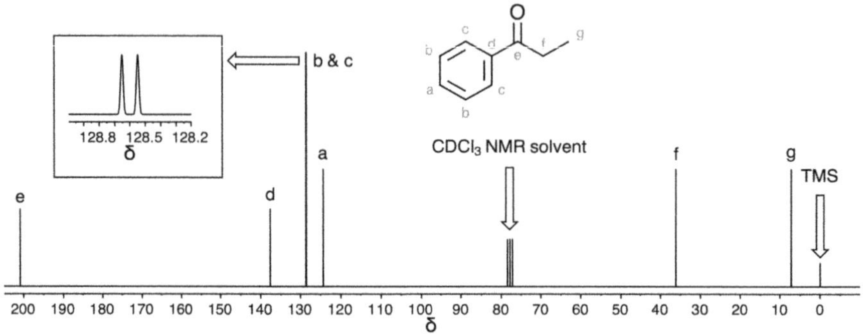

FIGURE 7.2 Carbon-13 NMR spectrum of propiophenone.

spectra, it is clear that two of the signals are much larger than the others. These two signals therefore correspond to carbons of type "b" and "c," since there are two carbons of each of these types due to the molecule's symmetry. Notice that these two peaks, while similar in chemical shift, are in fact distinct, as illustrated in the inset. As we noted earlier, overlap of signals for different types of carbon atoms is rare in ^{13}C NMR due to the large chemical shift range. The smaller signals in this chemical-shift region correlate with carbons "a" and "d." Due to resonance effects, aromatic rings can sometimes yield tricky chemical shift orderings; however, in this case, things work out fairly simply. Carbon "d" is closer to the ketone and is therefore more deshielded.

CHAPTER 7 PRACTICE PROBLEMS

1. Let's return to a few of the molecules that we've encountered in previous chapters. For each of the structures shown below (Figures 7.3–7.6), predict the number of signals in the carbon-13 NMR spectrum, as well as their chemical shift ranges.

   a.

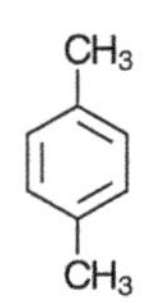

FIGURE 7.3 *Para*-xylene.

   b.

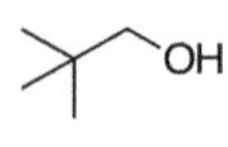

FIGURE 7.4 2,2-Dimethyl-1-propanol.

c.

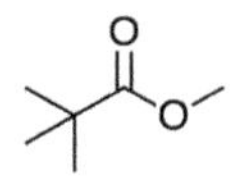

FIGURE 7.5 Methyl 2,2-dimethylpropionate.

d.

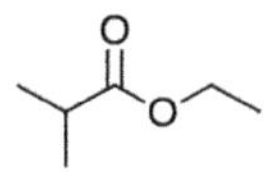

FIGURE 7.6 Ethyl 2-methylpropionate.

2. The structures of (*E*)-3-penten-2-one, (*E*)-3-penten-2-ol, and 2-pentanone are given in Figure 7.7, along with their carbon-13 NMR spectra. Match each structure with its corresponding spectrum.

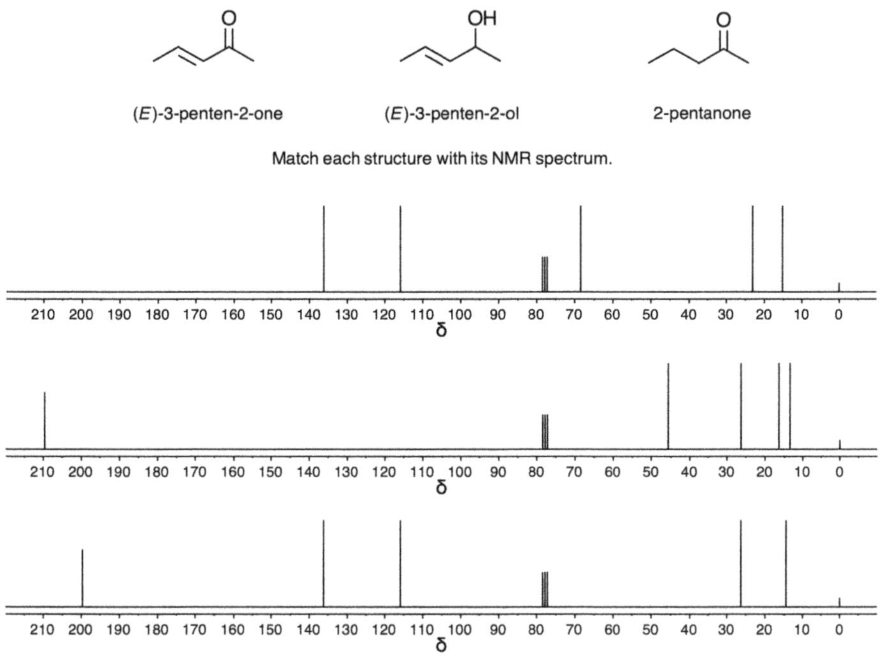

FIGURE 7.7 Three structures to be matched with their corresponding ^{13}C NMR spectra.

3. A solvolysis reaction of 3-bromo-2,2-dimethylbutane is shown in Figure 7.8. The mechanism of this reaction is S_N1, meaning that there's a carbocation intermediate. If there were no carbocation rearrangement, then the product would be 3,3-dimethyl-2-butanol. However, if carbocation rearrangement does occur, then the product will be 2,3-dimethyl-2-butanol.

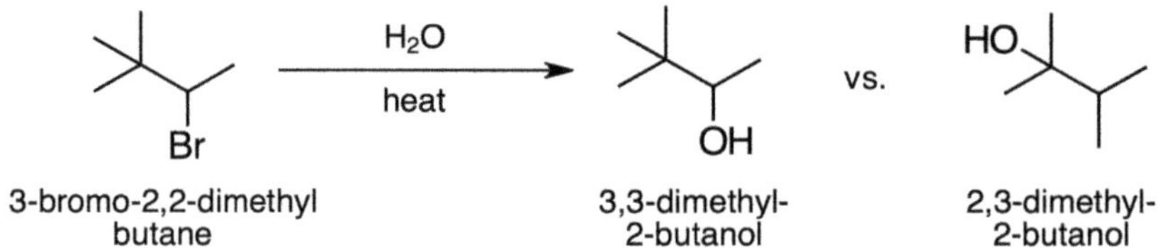

FIGURE 7.8 Solvolysis of 3-bromo-2,2-dimethylbutane.

Based on the carbon-13 NMR of the product (Figure 7.9), which compound did the solvolysis reaction produce?

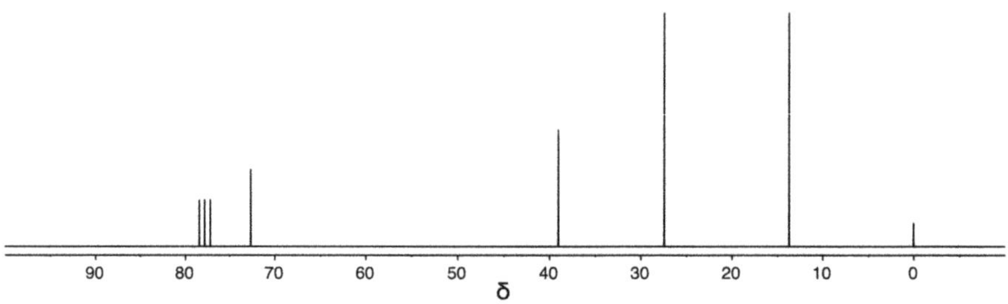

FIGURE 7.9 Carbon-13 NMR spectrum of solvolysis product.

4. There are four alcohols with the molecular formula $C_4H_{10}O$. Which one corresponds with the carbon-13 NMR spectrum in Figure 7.10?

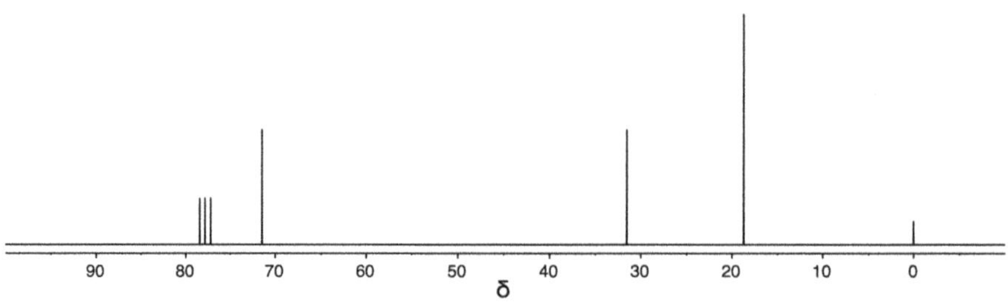

FIGURE 7.10 Carbon-13 NMR spectrum of an unknown alcohol ($C_4H_{10}O$).

SOLUTIONS TO CHAPTER 7 PRACTICE PROBLEMS

1.

a. The aromatic molecule *para*-xylene is a highly symmetrical molecule (Figure 7.11). Its two methyl groups (a) are identical, so their carbon atoms yield a single signal that should appear below 100 ppm in the ^{13}C NMR. In the aromatic ring itself, there are two types of carbon atoms: those bearing the methyl substituents (b) and those without any substituent (c). Therefore, there will only be two signals between 100 and 150 ppm. Since signal "c" accounts for four carbons while signal "b" accounts for only two carbons, we can safely say that "c" will be the larger signal in this region.

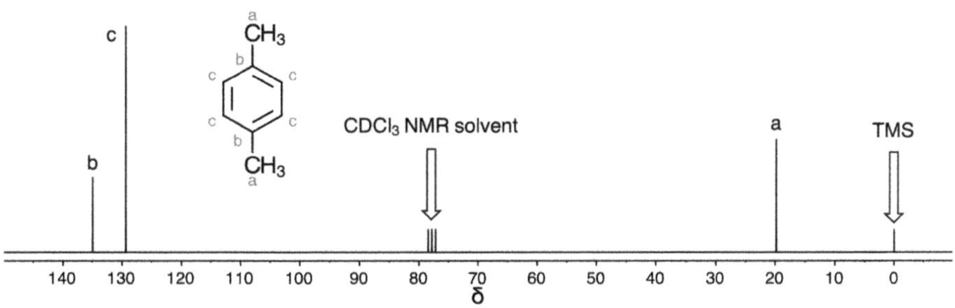

FIGURE 7.11 Carbon-13 NMR spectrum of *para*-xylene.

b. 2,2-Dimethyl-1-propanol contains three types of carbon atoms (Figure 7.12). The three methyl groups (a) of the *tert*-butyl group are identical because they all emanate from the same carbon (b). Carbon "b" is, of course, a different type of carbon atom, and carbon "c" is also unique. All three of these signals are for sp^3 hybridized carbons, so they all appear below 100 ppm, though we can say that carbon "c" will be the most deshielded since it is closest to the electron-withdrawing oxygen atom. We can also predict that signal "a" will not only be the most shielded since it is furthest from the functionality, but also that it will be the largest signal since it is caused by three identical carbon atoms.

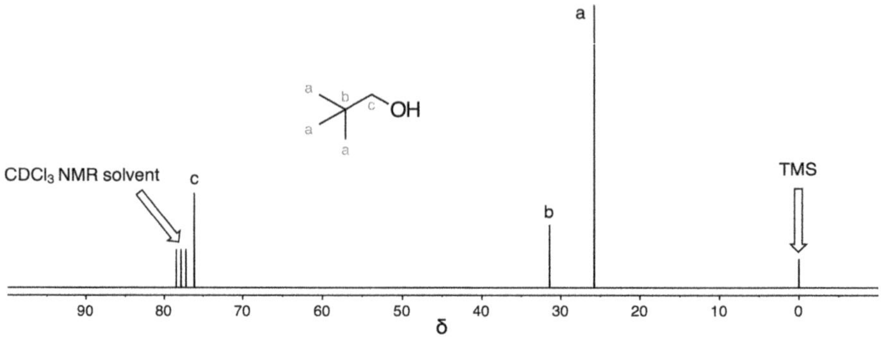

FIGURE 7.12 Carbon-13 NMR spectrum of 2,2-dimethyl-1-propanol.

c. Methyl 2,2-dimethylpropionate contains four types of carbon atoms (Figure 7.13). The carbonyl carbon (c) appears above 160 ppm. It is the only signal in this range. The remaining three types of carbon atoms (a, b, and d) are all sp^3 hybridized and therefore appear below 100 ppm. We can say that "d" will be the most deshielded because it is directly connected to an electronegative oxygen atom. The most shielded carbons will be type "a" since they are furthest from functionality. That signal will also be large because it is caused by three equivalent carbon atoms. The signal for "b" will be intermediate between "d" and "a" since it is adjacent to a carbonyl, which is deshielding but not as much so as the oxygen to which "d" is bonded.

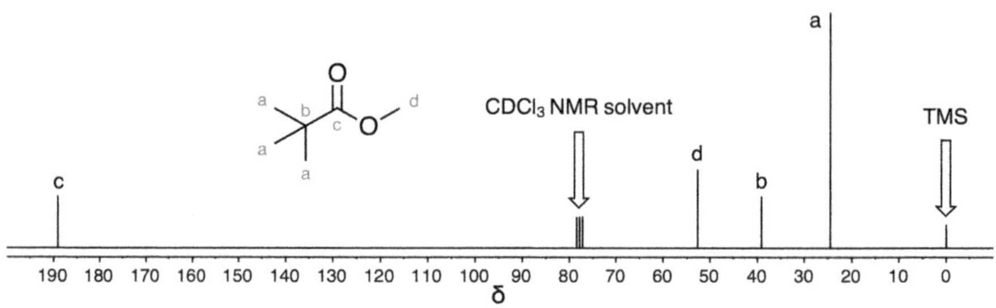

FIGURE 7.13 Carbon-13 NMR spectrum of methyl 2,2-dimethylpropionate.

d. Ethyl 2-methylpropionate consists of five types of carbon atoms (Figure 7.14). The most distinctive is the carbonyl carbon (c). It will appear above 160 ppm and is the only signal in this range. The remaining carbons all appear below 100 ppm. We can predict a couple of things about their ordering and relative size, though. Carbon "d" is directly bonded to oxygen, so it should be the most deshielded of the cluster below 100 ppm. Next comes carbon "b," which is adjacent to the less-deshielding carbonyl. Carbons of type "a" and "e" will produce the signals closest to 0 ppm, but the signal for the two equivalent carbons "a" will be larger than the signal for carbon "e."

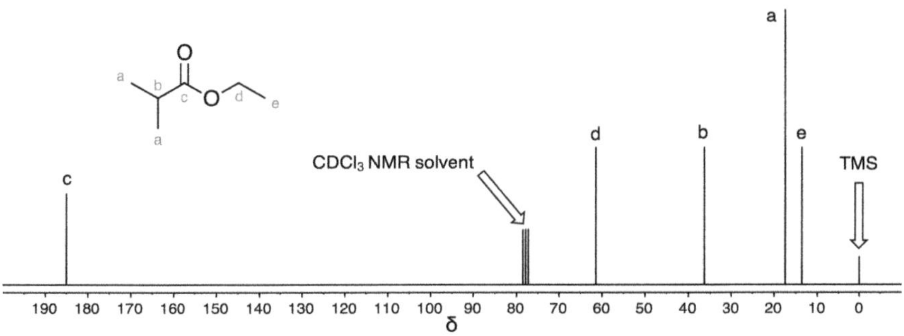

FIGURE 7.14 Carbon-13 NMR spectrum of ethyl 2-methylpropionate.

2. Let's assign the structures in the order in which they were presented. In each case, remember that the carbon atom of the $CDCl_3$ NMR solvent appears as three spikes around 77 ppm, and the carbons of TMS produce a signal at 0 ppm.

(*E*)-3-Penten-2-one (Figure 7.15) contains carbon atoms that will appear in all three chemical shift ranges shown in Figure 7.1. Carbons "a" and "e" are sp^3 hybridized and appear below 100 ppm. Carbons "b" and "c" are alkene carbons appearing between 100 and 150 ppm. Meanwhile, carbonyl carbon "d" appears above 160 ppm. Note that, in order to answer this question, you don't need to know exactly which peak below 100 ppm is "e" and which is "a," nor do you need to know exactly which signal from 100–150 ppm is "b" and which is "c."

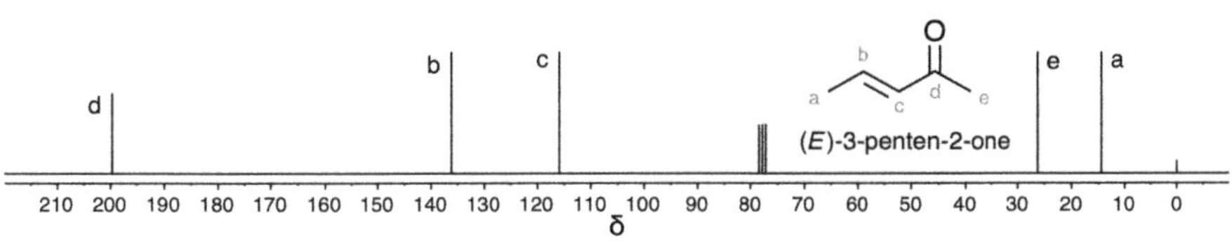

FIGURE 7.15 Carbon-13 NMR spectrum of (*E*)-3-penten-2-one.

(*E*)-3-Penten-2-ol contains carbon atoms of the sp^3 hybridized and alkene varieties. Therefore, two of the chemical shift regions in Figure 7.1 are occupied in its spectrum (Figure 7.16). Carbons "a," "d," and "e" all appear below 100 ppm. We know that methine "d" is the most deshielded of the trio because it is bonded to oxygen. The alkene carbons (b and c) appear from 100–150 ppm, and we don't need to know which is which for the sake of this question.

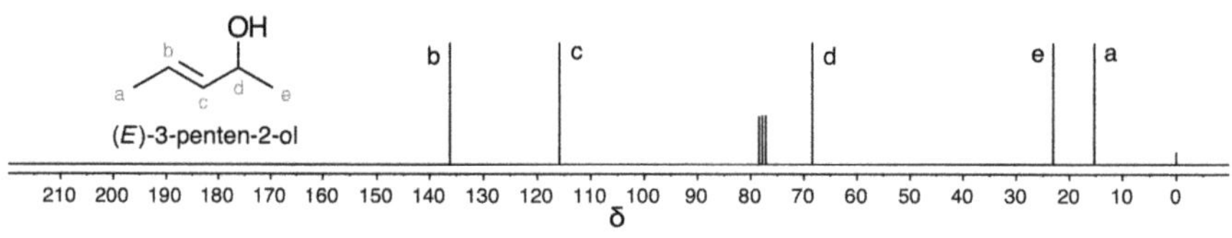

FIGURE 7.16 Carbon-13 NMR spectrum of (*E*)-3-penten-2-ol.

2-Pentanone also has carbons that will occupy only two of the three chemical shift regions; however, a different combination of regions is occupied in its spectrum (Figure 7.17). The carbonyl carbon (d) appears above 160 ppm. The remaining four carbons (a–c and e) fall below 100 ppm. Again, we don't need to specifically assign each carbon in this region in order to correctly match the molecules with their spectra.

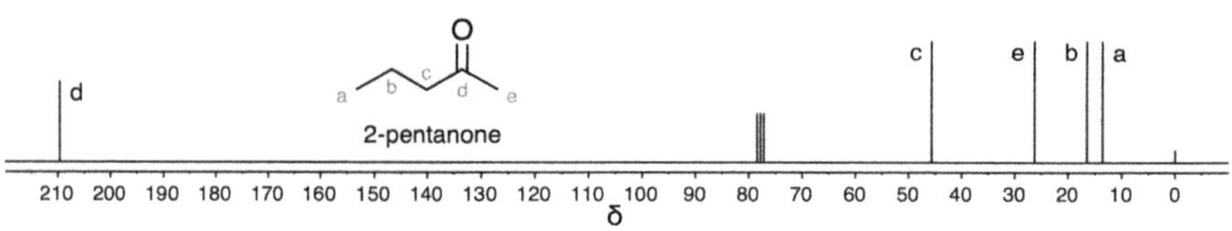

FIGURE 7.17 Carbon-13 NMR spectrum of 2-pentanone.

3. The structure of the reactant is not relevant when answering this question. Instead, we should consider the two possible product structures. It soon becomes apparent that both would yield four carbon-13 NMR signals, labeled a–d for 3,3-dimethyl-2-butanol and e–h for 2,3-dimethyl-2-butanol (Figure 7.18). In both cases, the four signals would fall below 100 ppm, so the chemical shift would also not differentiate.

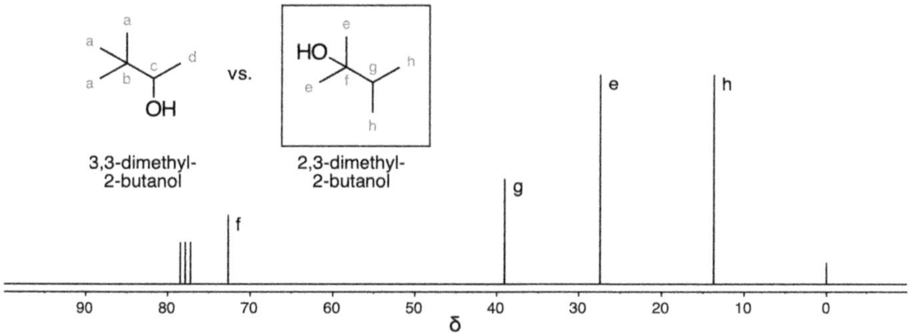

FIGURE 7.18 Carbon-13 NMR spectrum of 2,3-dimethyl-2-butanol.

What sets the two molecules apart is the number of carbon atoms giving rise to each signal. In the case of 3,3-dimethyl-2-butanol, one signal (a) is due to three carbon atoms, so it will be much larger than the other three signals (b–d), all of which are caused by a single carbon atom. To summarize, for 3,3-dimethyl-2-butanol, we expect one big signal and three smaller ones. On the other hand, 2,3-dimethyl-2-butanol has two signals resulting from two carbon atoms apiece (e and h) and two signals resulting from one carbon atom each (f and g). So, if it is the product, we anticipate a ^{13}C spectrum with two bigger signals and two smaller ones. That is, in fact, what we see in the spectrum provided, so 2,3-dimethyl-2-butanol is the product of this reaction.

4. There are four possible alcohol-containing isomers having the molecular formula $C_4H_{10}O$ (Figure 7.19). They include *n*-butyl alcohol, *sec*-butyl alcohol, isobutyl alcohol, and *tert*-butyl alcohol. Both *n*-butyl alcohol and *sec*-butyl alcohol would yield four carbon-13 NMR signals, while *tert*-butyl alcohol would give only two signals. Only isobutyl alcohol would produce three signals in the ^{13}C spectrum.

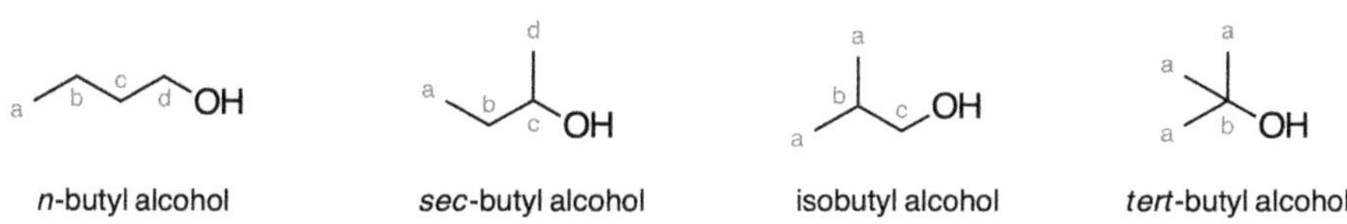

FIGURE 7.19 The four alcohol-containing isomers of $C_4H_{10}O$.

It is not necessary to assign the peaks to the carbons of isobutyl alcohol in order to answer this question, but Figure 7.20 shows these assignments. The most deshielded carbon (c) is bonded to the electron-withdrawing oxygen atom. As we move further away from the functionality, the signals have chemical shifts closer to 0 ppm.

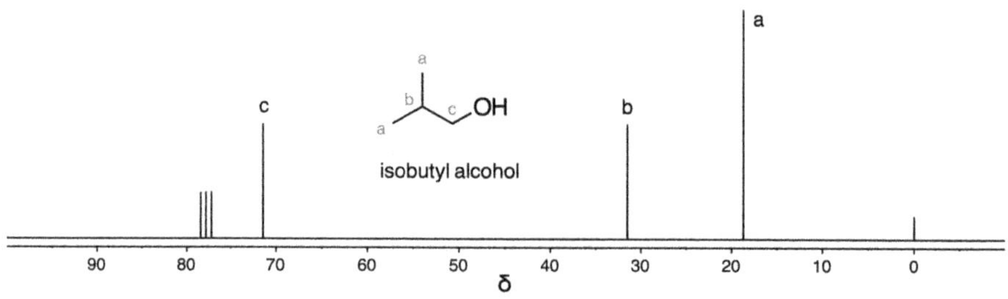

FIGURE 7.20 Carbon-13 NMR spectrum of isobutyl alcohol.

8 Practice Problems

1.

a. Assign the protons of 3-ethylacetophenone to the signals in its proton nuclear magnetic resonance (NMR) spectrum (Figure 8.1).

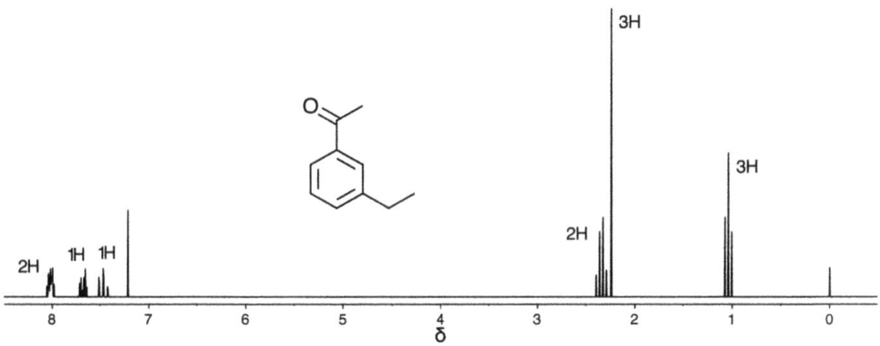

FIGURE 8.1 Proton NMR spectrum of 3-ethylacetophenone.

b. Assign the carbons of 3-ethylacetophenone to the signals in its ^{13}C NMR spectrum (Figure 8.2).

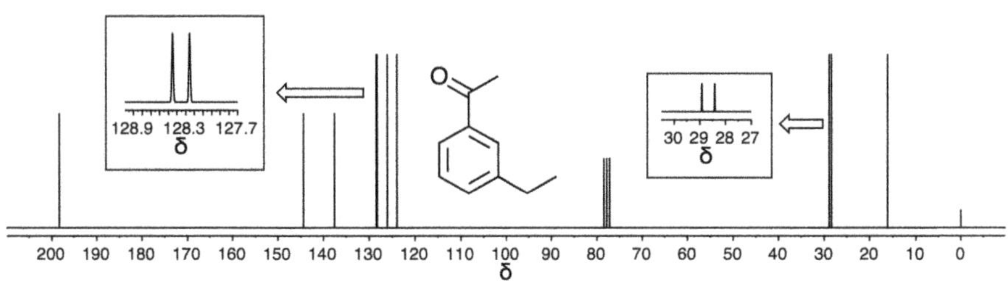

FIGURE 8.2 Carbon-13 NMR spectrum of 3-ethylacetophenone.

2.

a. Assign the protons of 1-pentanol to the signals in its proton NMR spectrum (Figure 8.3).

DOI: 10.1201/9781003783329-8

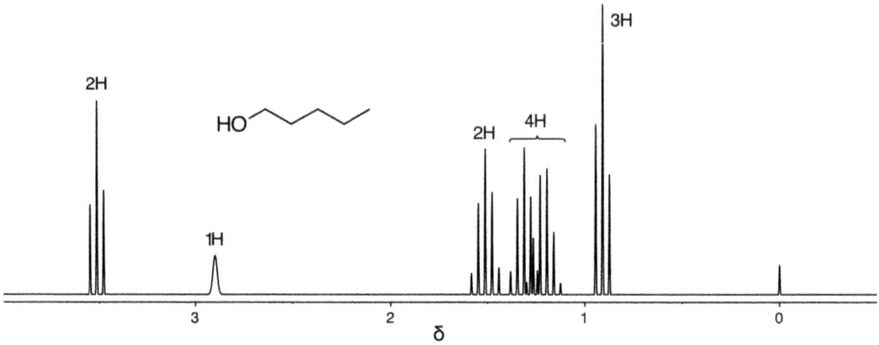

FIGURE 8.3 Proton NMR spectrum of 1-pentanol.

b. Assign the carbons of 1-pentanol to the signals in its ^{13}C NMR spectrum (Figure 8.4).

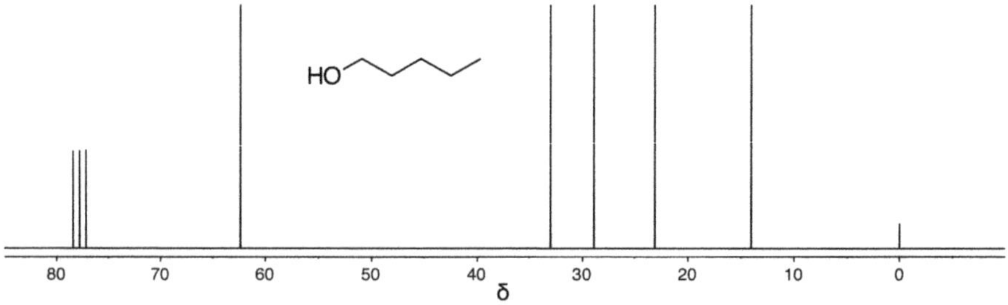

FIGURE 8.4 Carbon-13 NMR spectrum of 1-pentanol.

3.

a. Assign the protons of 2-iodopentane to the signals in its proton NMR spectrum (Figure 8.5).

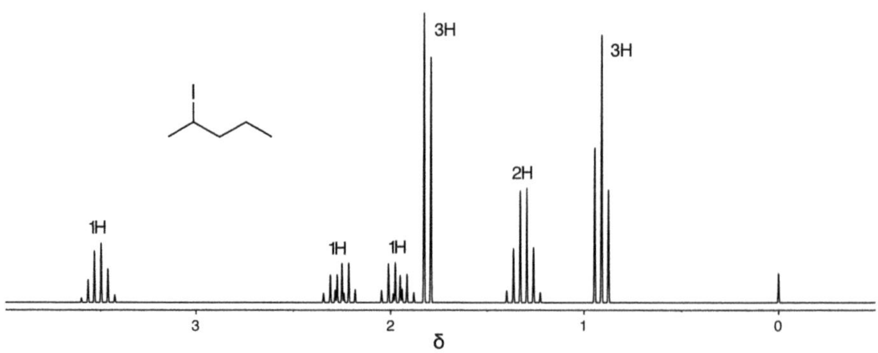

FIGURE 8.5 Proton NMR spectrum of 2-iodopentane.

b. Assign the carbons of 2-iodopentane to the signals in its ^{13}C NMR spectrum (Figure 8.6).

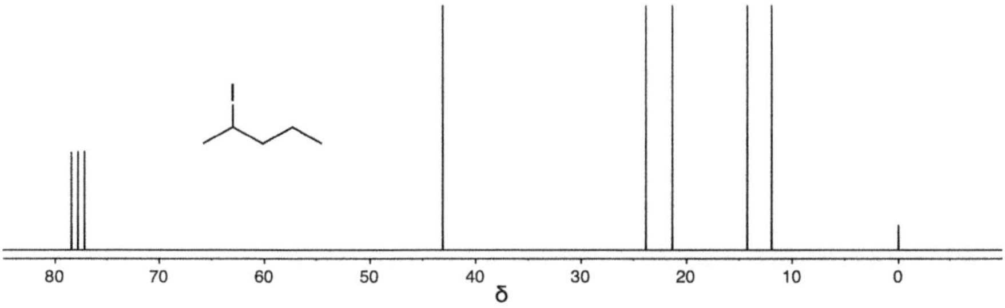

FIGURE 8.6 Carbon-13 NMR spectrum of 2-iodopentane.

4. The structures of benzyl methyl ether, ethyl phenyl ether, and phenethyl alcohol are given in Figures 8.7 and 8.8, along with their proton and carbon-13 NMR spectra, respectively. Match each structure with the corresponding spectrum.
   a. Match the structures with their proton NMR spectra.

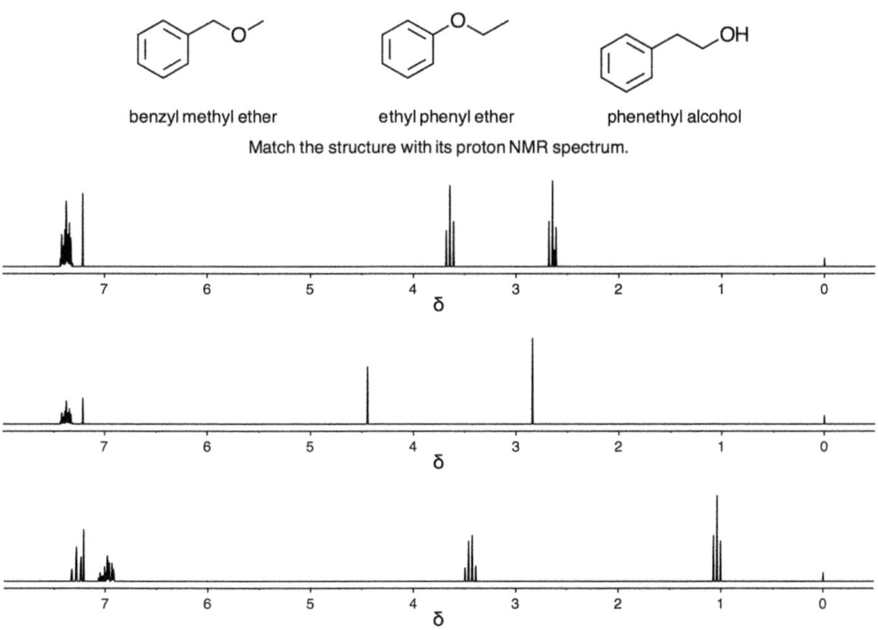

FIGURE 8.7 Three structures to be matched with their corresponding proton NMR spectra.

b. Match the structures with their carbon-13 NMR spectra. Note that only the region from roughly 0 to 100 ppm has been shown in each case. Base your analysis on differences in this segment of the spectra.

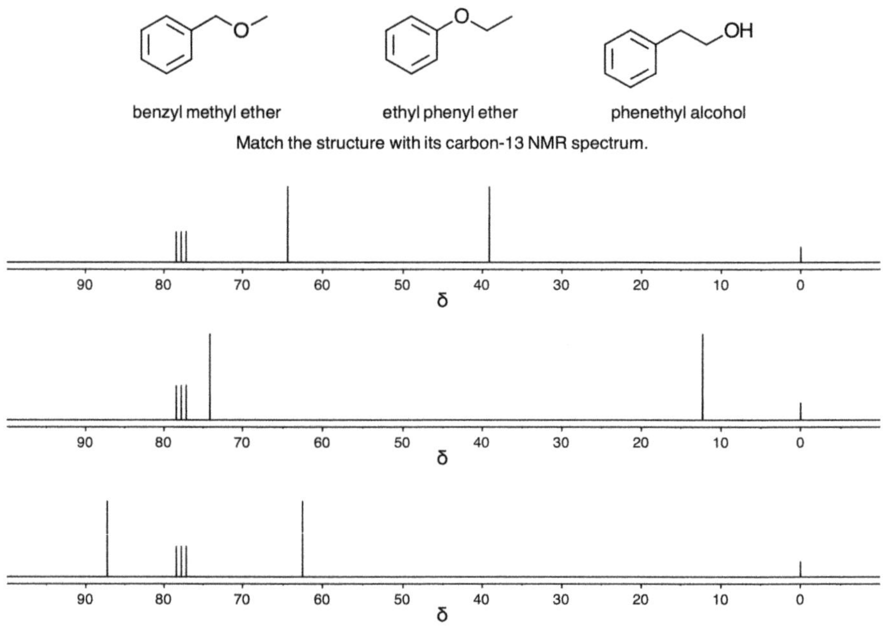

FIGURE 8.8 Three structures to be matched with their corresponding carbon-13 NMR spectra.

5. The structures of *para*-cresol, 4-*tert*-butylphenol, and 2-phenyl-2-propanol are given in Figure 8.9, along with their proton NMR spectra containing integral traces. Match each structure with the corresponding spectrum.

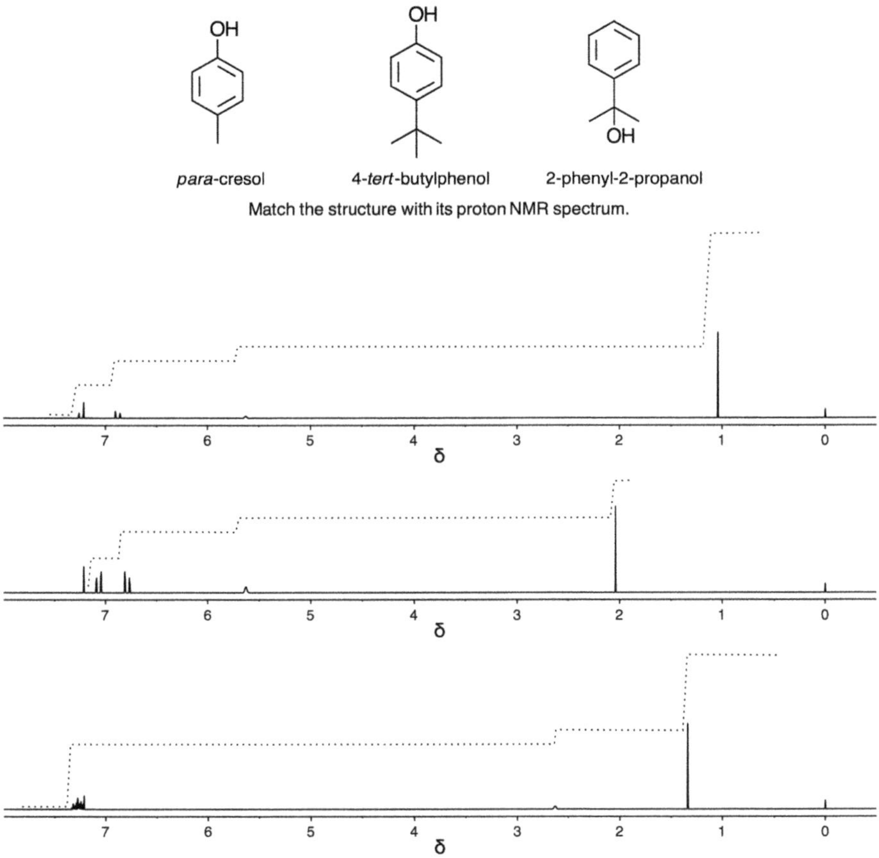

FIGURE 8.9 Three structures to be matched with their corresponding proton NMR spectra.

6. Three structures are provided in Figure 8.10: 1-phenyl-2-propanone, propiophenone, and 4-methylacetophenone. One of these three compounds produces both the carbon-13 and proton NMR spectra shown. Which compound is it?

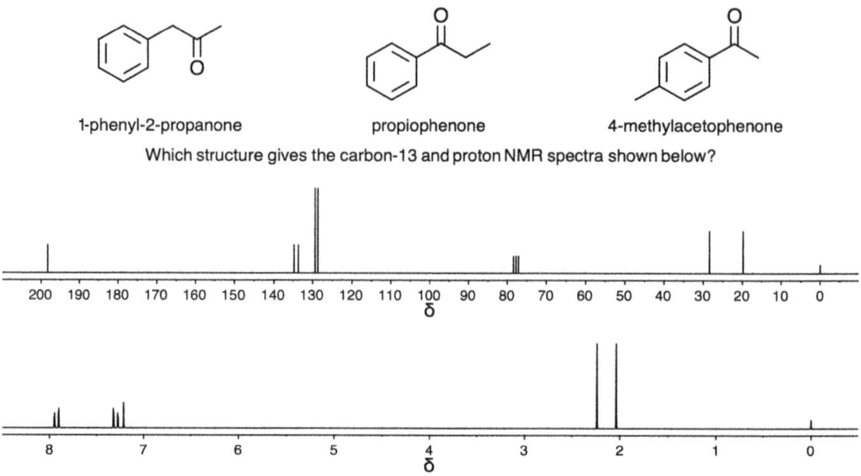

FIGURE 8.10 Three structures, one of which produces the carbon-13 and proton NMR spectra shown.

7. Draw a splitting tree for the proton labeled "a" in the structure of (*S*)-2-isobutyloxirane (Figure 8.11). Be sure to think carefully about the effect of the chiral center (c). What type of splitting pattern does proton "a" produce in the proton NMR spectrum?

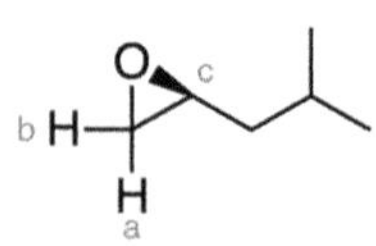

FIGURE 8.11 (*S*)-2-Isobutyloxirane.

8. Due to conformational constraint, aldehyde protons often don't experience visible coupling with their vicinal neighbors, and if they do, the coupling constant (*J* value) is very small. Assuming that the aldehyde proton will not couple or will couple only minimally with its neighbors, predict the appearance of the proton NMR spectrum of (3-chlorophenyl)acetaldehyde (Figure 8.12).

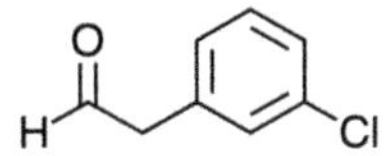

FIGURE 8.12 (3-Chlorophenyl)acetaldehyde.

9. Use a resonance-based argument to explain the extraordinary deshielding of one of the vinyl protons of crotonaldehyde (Figure 8.13).

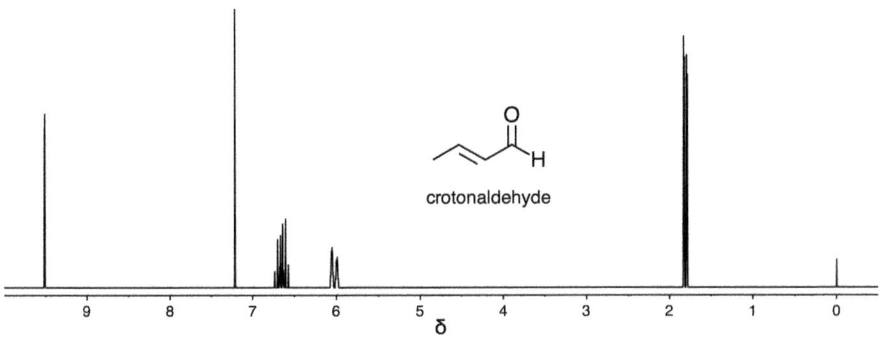

FIGURE 8.13 Proton NMR spectrum of crotonaldehyde.

10. In the proton NMR spectrum of toluene, the aryl proton signals overlap to form a messy multiplet (Figure 8.14). However, in the proton NMR spectrum of acetophenone, they resolve into three distinct signals. Explain this difference.

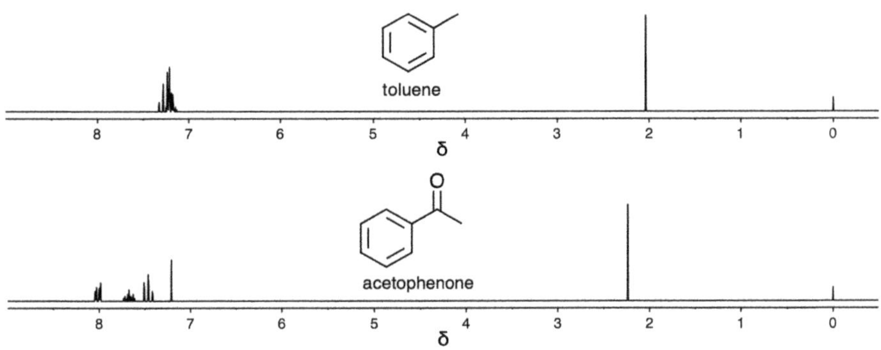

FIGURE 8.14 Proton NMR spectra of toluene and acetophenone.

11. An unknown molecule, compound A, has the molecular formula $C_5H_{10}O$ and produces the ^{13}C NMR spectrum in Figure 8.15. What is the structure of compound A?

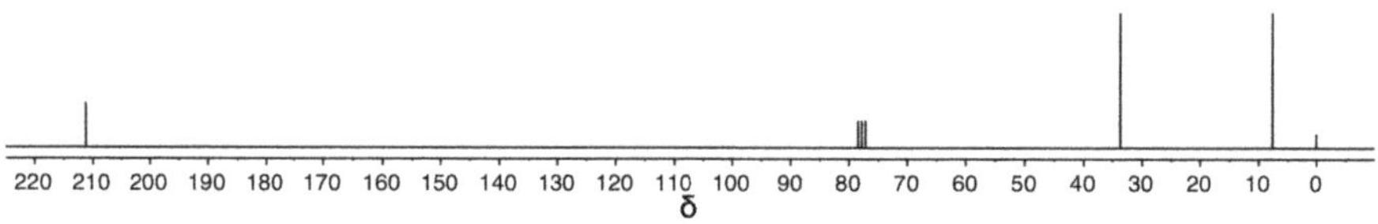

FIGURE 8.15 Carbon-13 NMR spectrum of an unknown molecule with the formula $C_5H_{10}O$.

12. An unknown substance, compound B, has the molecular formula C_8H_{10}. Compound B's carbon-13 NMR spectrum is given in Figure 8.16. Identify the structure of compound B?

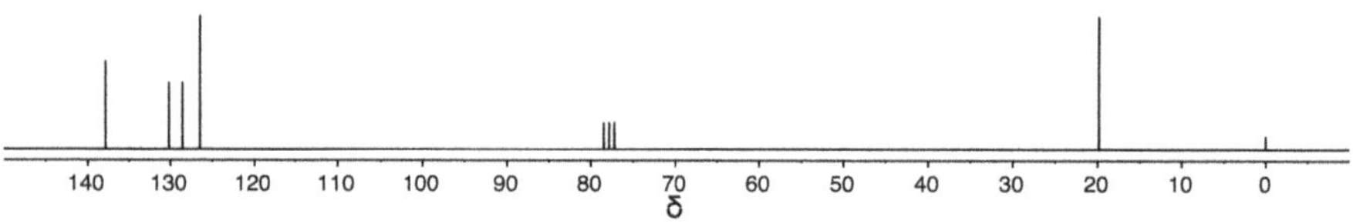

FIGURE 8.16 Carbon-13 NMR spectrum of an unknown molecule with the formula C_8H_{10}.

13. The proton NMR spectrum of compound C, an unknown alkyl halide having the molecular formula $C_6H_{13}Cl$, is given in Figure 8.17. What is the structure of compound C?

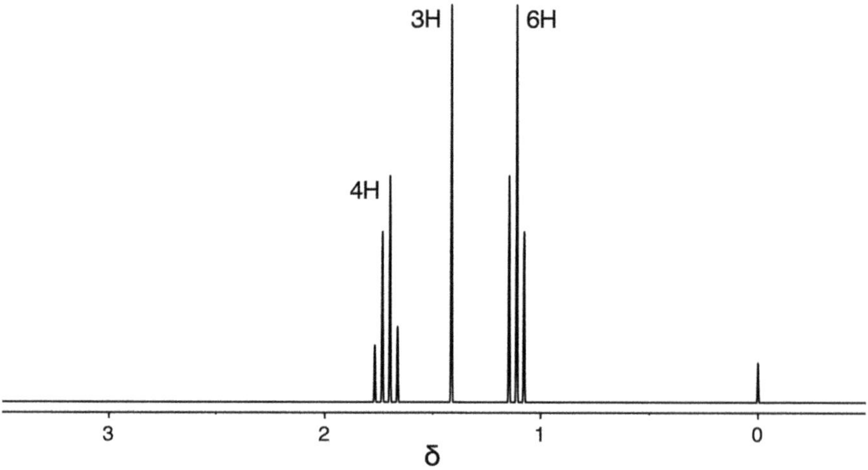

FIGURE 8.17 Proton NMR spectrum of an unknown alkyl halide with the formula $C_6H_{13}Cl$.

14. Compound D is an unknown having the formula C_8H_8O. Both its proton and carbon-13 NMR spectra are given in Figure 8.18. Identify compound D's structure.

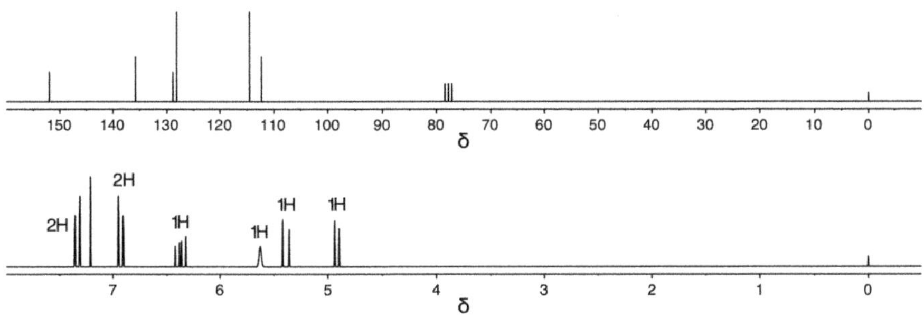

FIGURE 8.18 Carbon-13 and proton NMR spectra of compound D (C_8H_8O).

15. Figure 8.19 shows the carbon-13 and proton NMR spectra of an unknown molecule, compound E. Compound E has the molecular formula $C_9H_{10}O_2$. What is compound E's structure?

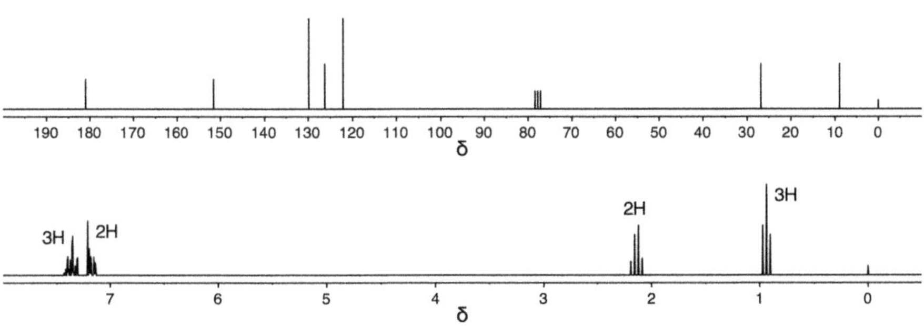

FIGURE 8.19 Carbon-13 and proton NMR spectra of compound E ($C_9H_{10}O_2$).

16. Compound F has the molecular formula $C_6H_{13}NO$ and produces the spectra shown in Figure 8.20. Identify the structure of compound F.

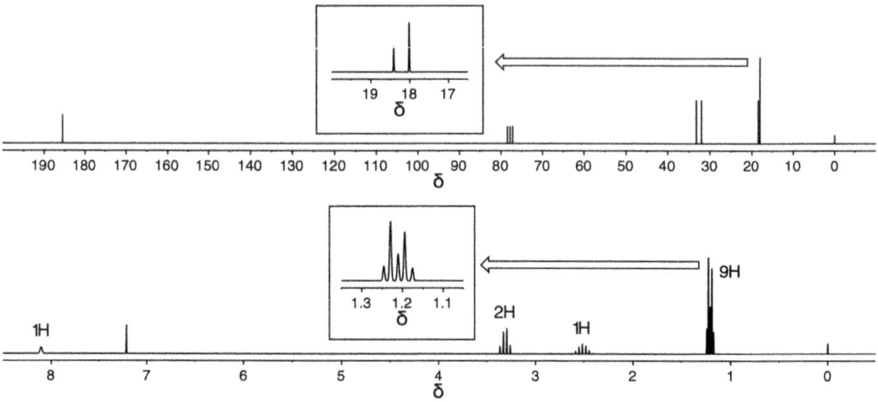

FIGURE 8.20 Carbon-13 and proton NMR spectra of compound F ($C_6H_{13}NO$).

17. Compound G is a substance of unknown molecular structure having the formula $C_7H_{16}O$. Its spectra appear in Figure 8.21. What is the identity of compound G?

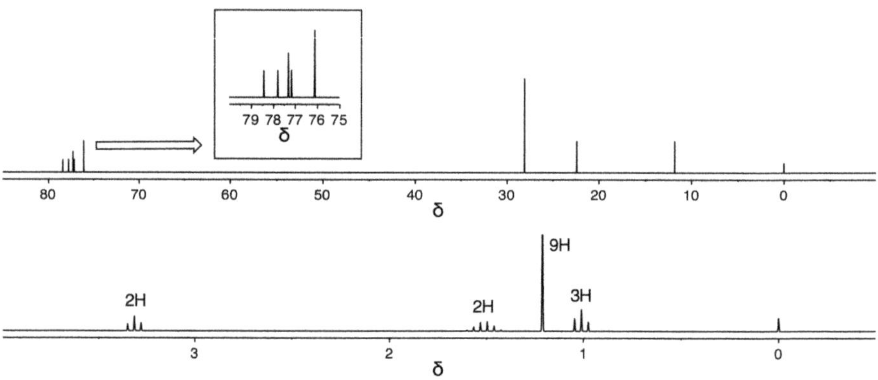

FIGURE 8.21 Carbon-13 and proton NMR spectra of compound G ($C_7H_{16}O$).

18. Compound H has the molecular formula $C_{11}H_{14}O_2$ and produces the spectra shown in Figure 8.22. What is its structure?

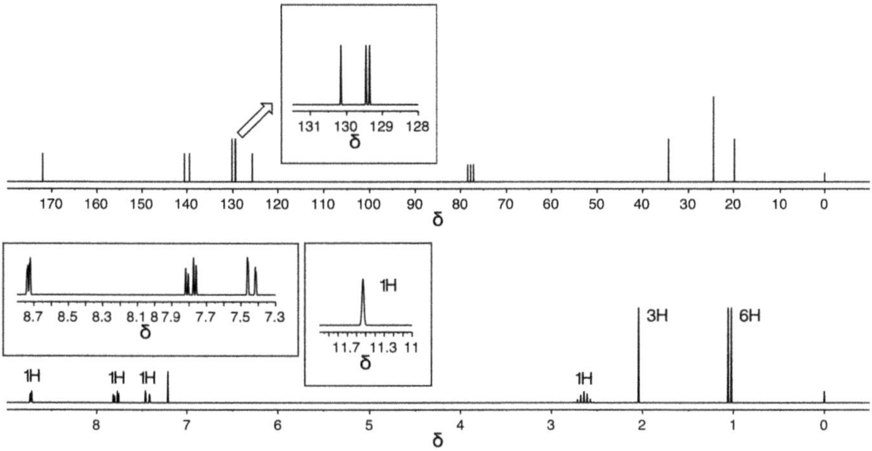

FIGURE 8.22 Carbon-13 and proton NMR spectra of compound H ($C_{11}H_{14}O_2$).

19. An unknown molecule with the formula $C_7H_{17}N$, compound I, produces the NMR spectra in Figure 8.23. What is the structure of compound I?

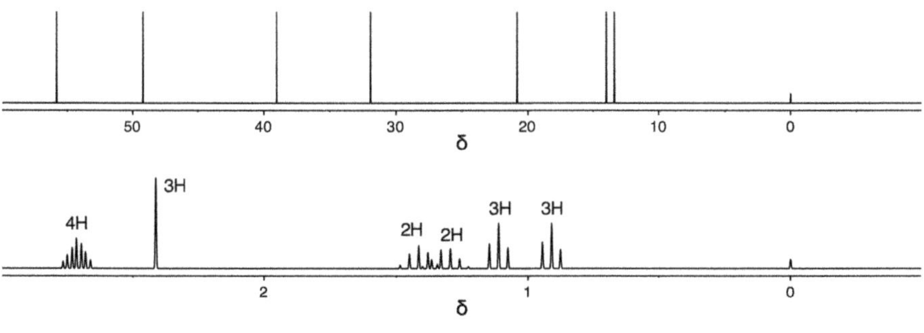

FIGURE 8.23 Carbon-13 and proton NMR spectra of compound I ($C_7H_{17}N$).

20. Compound J has the molecular formula $C_{11}H_{14}O$ and the NMR spectra shown in Figure 8.24. What is its structure?

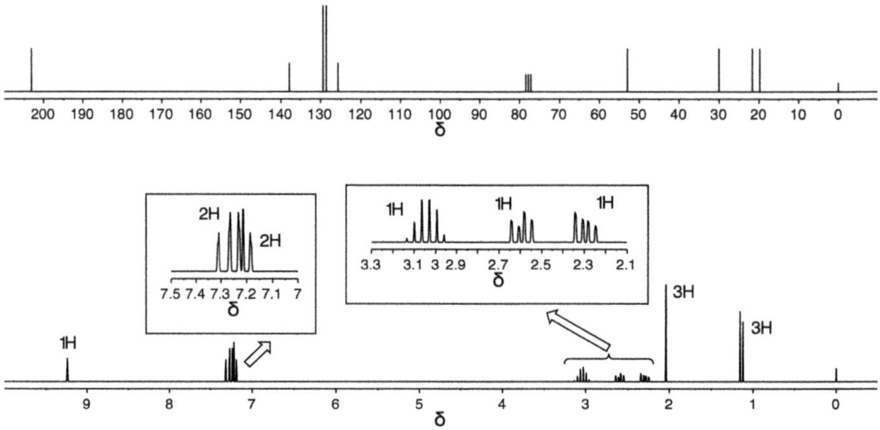

FIGURE 8.24 Carbon-13 and proton NMR spectra of compound J ($C_{11}H_{14}O$).

9 Solutions to Practice Problems

1.

a. There are seven types of protons in 3-ethylacetophenone (a–g, Figure 9.1). Let's address the methyl and methylene groups first. Methyl group "g" is furthest from functionality, so it should produce the most shielded signal (i.e., the one closest to 0 ppm, not counting the internal standard, TMS). The triplet at about 1 ppm fits the bill, and its integration (3H) as well as splitting (triplet due to two neighbors) reinforce our assignment. Both methyl "a" and methylene "f" are adjacent to a double bond, so they will both appear between 1.5 and 2.5 ppm. However, they are readily differentiated by their integration and splitting. Methyl "a" integrates for three hydrogens and is a singlet because it has no neighbors. Methylene "f" integrates for two hydrogens and is a quartet owing to its three neighbors (g).

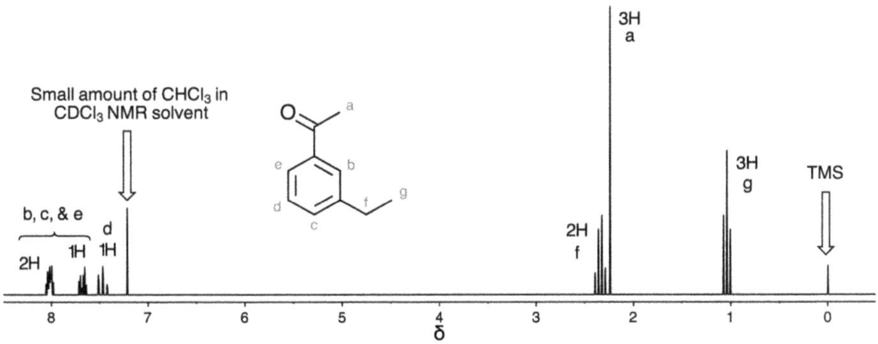

FIGURE 9.1 A labeled proton NMR spectrum of 3-ethylacetophenone.

Now, let's turn our attention to the aromatic ring. Distinguishing the signals in this region is trickier. Two of the proton signals overlap, giving a messy multiplet that integrates for two hydrogens. Another signal is a complex multiplet. This entire grouping can be attributed to protons "b," "c," and "e." The signal at about 7.5 ppm is a reasonably clear apparent triplet. That would most closely match proton "d," which has two *ortho* neighbors and therefore is split by two fairly large *J*-value couplings. Though the *J* values for coupling with "c" and "e" will differ, they may not differ by much, causing the middle spikes in the second coupling to overlap, yielding an apparent triplet.

DOI: 10.1201/9781003783329-9

b. There are ten types of carbon atoms in 3-ethylacetophenone (a–j, Figure 9.2). These ten signals will be split across the three main chemical shift regions found in ^{13}C nuclear magnetic resonance (NMR). The carbonyl carbon (b) appears in the region above 160 ppm (specifically near 200 ppm). The six aromatic carbons appear between 100 and 150 ppm. The inset shows that two signals, though close in chemical shift, are distinct. Notice that two signals in this region are a bit shorter than the others. Carbons with no attached hydrogens tend to cause less intense signals than carbon atoms bearing hydrogen, so we can surmise that the shorter signals are due to carbons "c" and "e." The taller signals in this region are the remaining aromatic carbons. Finally, the sp^3 hybridized carbons (a, i, and j) fall below 100 ppm. Carbons "a" and "i" are similarly deshielded and therefore not easily distinguished. However, carbon "j" is furthest from functionality, therefore giving the most shielded signal (at about 15 ppm).

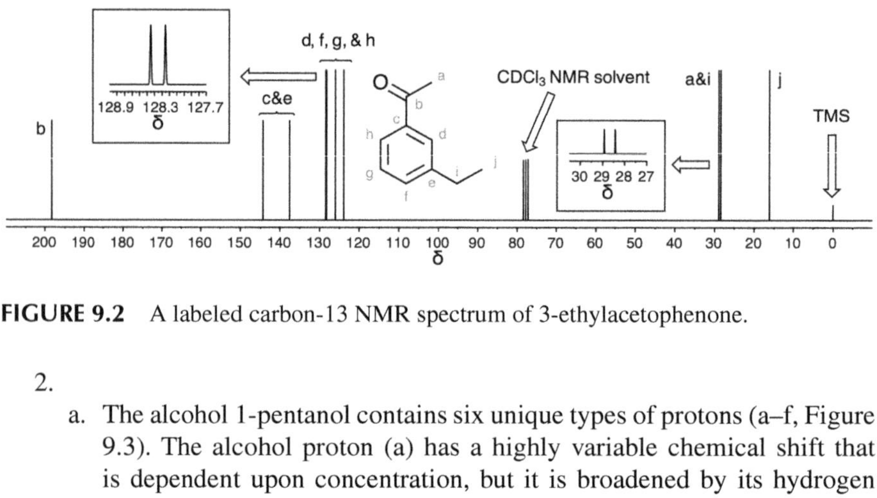

FIGURE 9.2 A labeled carbon-13 NMR spectrum of 3-ethylacetophenone.

2.

a. The alcohol 1-pentanol contains six unique types of protons (a–f, Figure 9.3). The alcohol proton (a) has a highly variable chemical shift that is dependent upon concentration, but it is broadened by its hydrogen bonding, which makes it quite easy to spot. Methylene protons "b" are of the type H–C–O, which puts them in the chemical shift range of 2.5–4.5 ppm. There is one remaining signal in this region. It integrates for two hydrogens and is split into a triplet by its two neighbors (c).

As we move progressively further away from the functional group, we expect the chemical shift of each signal to be incrementally more shielded (i.e., closer to 0 ppm). Methylene protons "c" therefore yield the next signal. Once again, their signal integrates for two hydrogens. These protons have a total of four neighbors (two of type "b" and two of type "d"). Since the neighbors are chemically inequivalent, we could see complex splitting. However, in this case, they are similar enough to one another that the n + 1 rule holds, making signal "c" a quintet.

Methylenes "d" and "e" are the next signals we encounter. They are in similar enough chemical environments that their signals overlap a bit and are integrated together for a total of four hydrogens. When signals overlap, it is often not easy to distinguish their individual splitting patterns.

Lastly, the methyl group (f) yields the most shielded signal (except for the internal standard, TMS). It integrates for three hydrogens and is split into a triplet by its two neighbors (e).

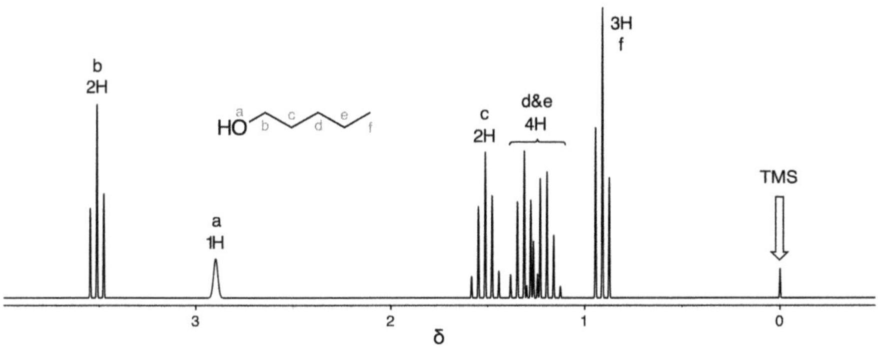

FIGURE 9.3 A labeled proton NMR spectrum of 1-pentanol.

b. 1-Pentanol contains five types of carbons. All are sp^3 hybridized and therefore appear below 100 ppm. However, we can still differentiate them based on chemical shift differences. Generally speaking, as we move further from a functional group, the signals become more shielded (i.e., closer to 0 ppm). That trend is observed here. The most deshielded signal (not counting the solvent) is for carbon "a," which is closest to the alcohol. Then, we successively encounter the signals for carbons "b" through "e" as we move along the x-axis, getting closer to 0 ppm with each step further away from the alcohol (Figure 9.4).

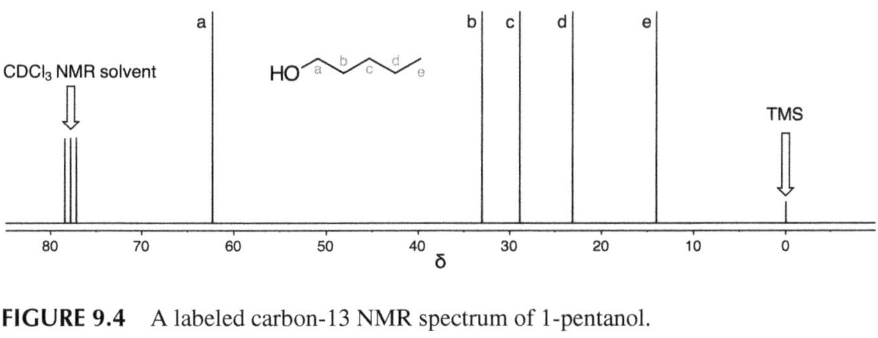

FIGURE 9.4 A labeled carbon-13 NMR spectrum of 1-pentanol.

3.

a. 2-Iodopentane has six types of protons (a–f, Figure 9.5). The methine proton (b) is the most deshielded since it is of the type H–C–I, putting it in the chemical shift range of 2.5–4.5 ppm. It integrates for one hydrogen. It also has five neighbors of three different types (a, c, and d). Since they are of different types, we could see complex splitting if their *J* values were significantly different. However, in this case, their *J* values are not too different, so the n + 1 rule holds, making the signal a sextet.

The methylene protons "c and d" are diastereotopic since there is a chiral center (b) in the molecule. This means that each of these methylene protons will yield its own signal integrating for one hydrogen. These signals are fairly deshielded since they are still near the functional group. Their splitting patterns are complex because they each have two neighbors of type "e," a methine neighbor (b), and they split each other as geminal neighbors since they are diastereotopic.

Methyl protons "a" are relatively deshielded due to proximity to the functional group. Their signal integrates for three hydrogens and is split into a doublet by their sole neighbor (b).

The methylene protons "e" are also formally diastereotopic, like the methylene protons "c and d." However, given the greater distance from the chiral center, the two protons of type "e" are not different enough to actually yield distinct signals. We see them as a single signal integrating for two hydrogens. They have five neighbors of multiple types (c, d, and f), yet the n + 1 rule has held here, yielding a sextet, because the *J* values of the neighbors end up not being all that different.

Finally, methyl group "f" yields a shielded triplet (due to its two neighbors "e") integrating for three hydrogens.

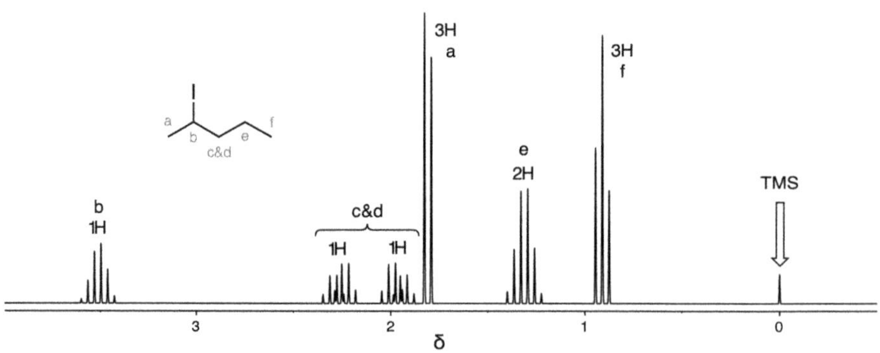

FIGURE 9.5 A labeled proton NMR spectrum of 2-iodopentane.

b. 2-Iodopentane possesses five unique types of carbon (a–e, Figure 9.6). Carbon "b" is expected to be the most deshielded since it is directly bonded to the electron-withdrawing halogen. Carbons "a" and "c" are still relatively close to the functional group, so they will be somewhat deshielded. Their exact chemical-shift ordering may not be obvious, so we can group them together. Lastly, carbons "d" and "e" are the most distant from the functional group and therefore the most shielded. Since "e" is further from iodine than "d," it is the signal closest to 0 ppm (not counting the internal standard, TMS).

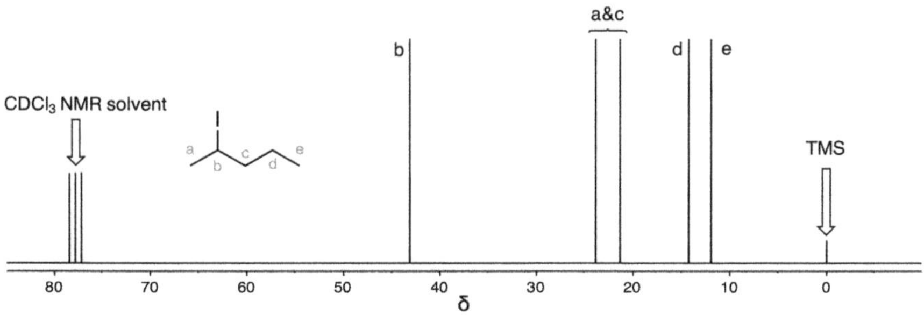

FIGURE 9.6 A labeled carbon-13 NMR spectrum of 2-iodopentane.

4.

a. The quickest and easiest way to complete the matching in this problem is to exploit the clear difference in the multiplicity of the more shielded signals. For each compound, the aromatic region consists of messy and overlapping signals. That is of little use in differentiating the spectra. Even if they were clearer, each compound contains a monosubstituted aromatic ring, so there is not a great difference anyway.

However, for the methyl and/or methylene groups of each compound, there's a pronounced difference in the expected splitting pattern. The first spectrum contains two triplets between 2.5 and 4.5 ppm (Figure 9.7). Phenethyl alcohol is expected to display triplets (t) for each methylene group. Remember that hydrogen-bonding protons, like that of the alcohol, do not typically participate in splitting, unless the sample is extraordinarily dilute.

The second spectrum has two singlets (s) in the same chemical shift region (2.5–4.5 ppm). Only benzyl methyl ether will give that pattern for its methyl and methylene groups.

The final spectrum contains a quartet (q) between 2.5 and 4.5 ppm and a triplet (t) between 0 and 1.5 ppm. This is consistent with the expected multiplicity of the methylene and methyl groups in its structure, respectively.

In case you were wondering about the hydroxyl (OH) proton of phenethyl alcohol, it actually happens to overlap with the triplet near 2.6 ppm in this spectrum, as shown in the inset in Figure 9.7.

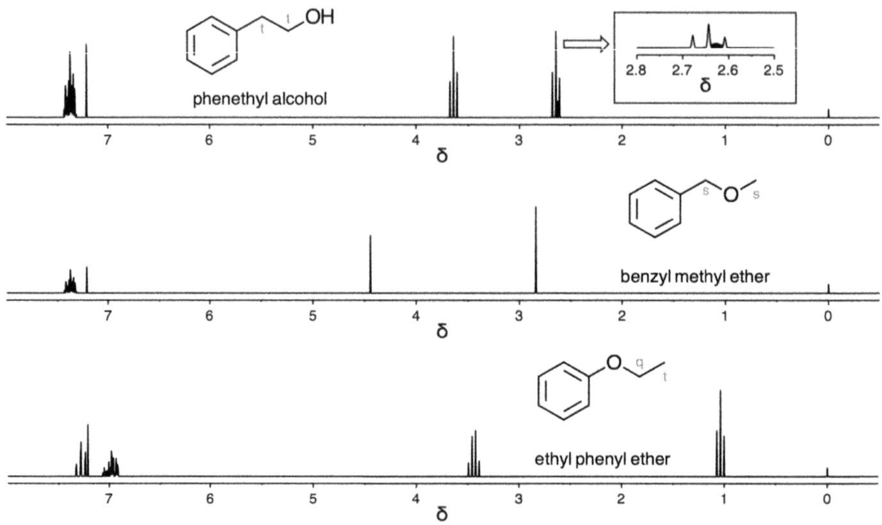

FIGURE 9.7 Three structures matched with their corresponding proton NMR spectra.

b. The aromatic region (100–150 ppm) is not shown for any of these three carbon-13 spectra. It would not be very helpful in drawing a distinction because, in each case, we'd expect four signals for the aromatic ring due to its symmetry. Though there might be some chemical shift variations, the more clear and striking differences are found in the region below 100 ppm, where the sp^3 hybridized carbons are found.

One of the sp^3 carbons (b) of phenethyl alcohol is highly deshielded because it is next to an oxygen atom, while the other (a) is only moderately deshielded since it is adjacent to a less electron-withdrawing aromatic ring.

Ethyl phenyl ether has one highly deshielded sp^3 carbon (c) that is adjacent to oxygen, and one shielded sp^3 carbon (d) that is as far removed as possible from the functionality.

Benzyl methyl ether has two highly deshielded sp^3 carbons. One (f) is adjacent to oxygen, and the other (e) is even more deshielded because it is between oxygen and the aromatic ring (Figure 9.8).

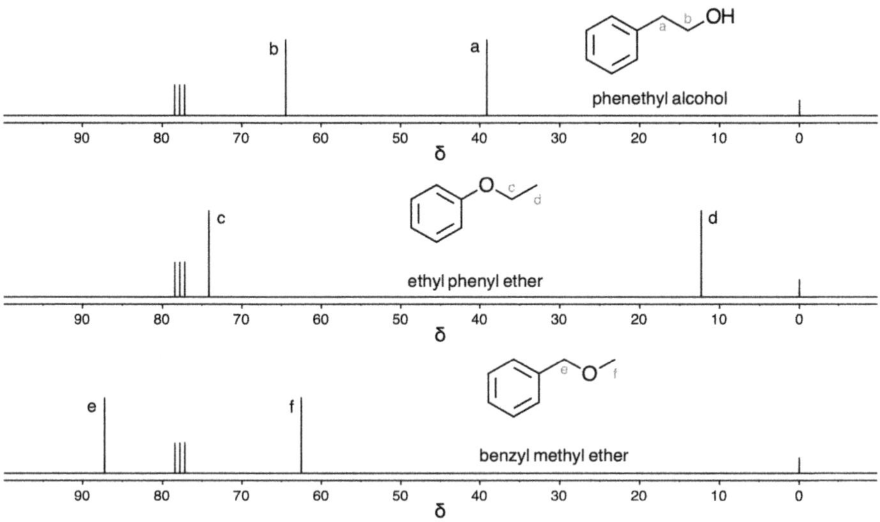

FIGURE 9.8 Three structures matched with their corresponding carbon-13 NMR spectra.

5. A prominent difference between the spectra is found in their integral traces. Let's examine the spectra in the order in which they were presented. The first spectrum has an approximate ratio of integrals of 2:2:1:9 (Figure 9.9). Recall that this is obtained by measuring each vertical height change in the integral trace and subsequently dividing all the values by the smallest one. Only 4-*tert*-butylphenol has a group of protons (the three equivalent methyls, d) that would integrate for nine.

The second spectrum has an approximate integral ratio of 2:2:1:3. Only *para*-cresol has a unique, single methyl group (d) that would yield a signal integrating for three hydrogens.

The third spectrum has an approximate integral ratio of 5:1:6. Only 2-phenyl-2-propanol has two identical methyl groups (d) that would produce a signal integrating for six protons.

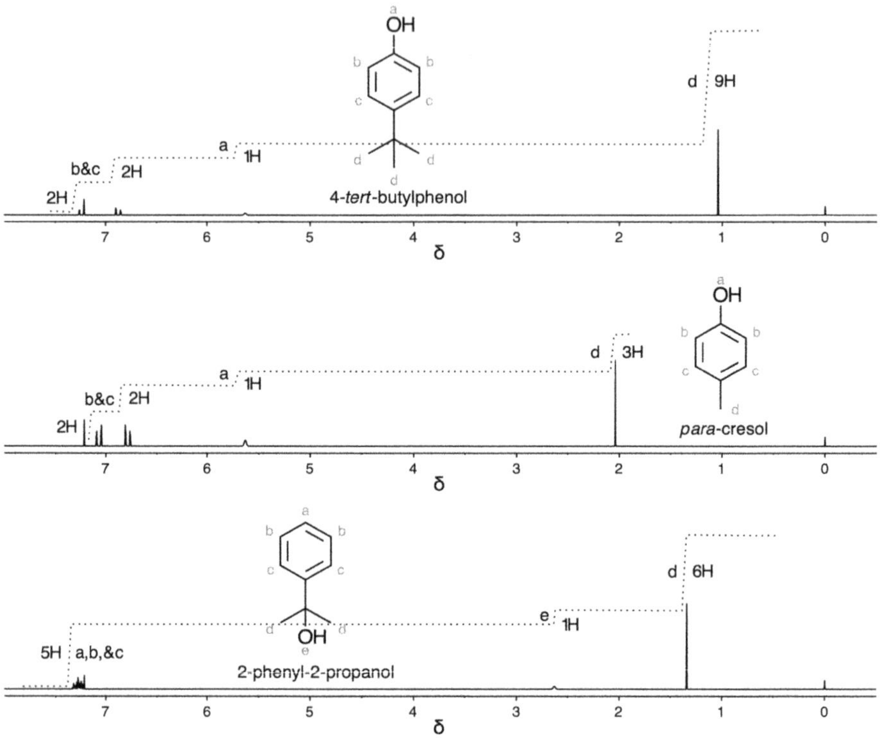

FIGURE 9.9 Three structures matched with their corresponding proton NMR spectra.

6. There are various differences in the anticipated spectra of the three molecules given. The best problem-solving strategy is to look for the clearest, most obvious distinctions. For instance, we can quickly rule out propiophenone because its ethyl group would yield two signals having splitting (specifically a quartet and triplet) below 2.5 ppm.

That facet of the anticipated spectra does not distinguish 1-phenyl-2-propanone from 4-methylacetophenone as easily, though. Both would be expected to generate two singlets in this general chemical shift region, so we must dig more deeply to differentiate them.

The aromatic region of the proton NMR (6.5–8 ppm) shows two doublets. 4-Methylacetophenone would give that pattern. There are two types of aryl protons in its structure (c and d), and each is split into a doublet by its single vicinal neighbor. In contrast, 1-phenyl-2-propanone would have a third signal in this region, and the splitting patterns would be more complex as well. Therefore, we've identified the spectra as resulting from 4-methylacetophenone (Figure 9.10).

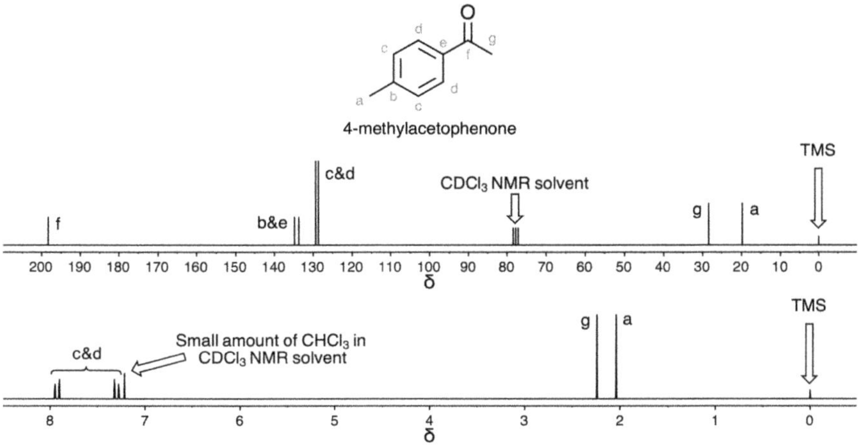

FIGURE 9.10 4-Methylacetophenone produces the carbon-13 and proton NMR spectra shown.

7. Due to the adjacent chiral center (c), the methylene protons (a and b) of the epoxide are diastereotopic, meaning that they are chemically inequivalent and cause different signals (Figure 9.11). As such, they will act as geminal neighbors of one another.

Therefore, proton "a" is split into a doublet by its geminal neighbor (b) with a coupling constant of about 12–15 Hz. Then, each of those two spikes is split into two by the coupling of "a" with its vicinal neighbor (c). This second splitting event has a smaller *J* value (6–8 Hz). The result is four spikes of roughly equivalent height, which is called a doublet of doublets.

The NMR signal is shown. It experiences a bit of second-order coupling, or roofing, due to proximity to another nearby signal (on the left, but omitted from the diagram) with which it couples.

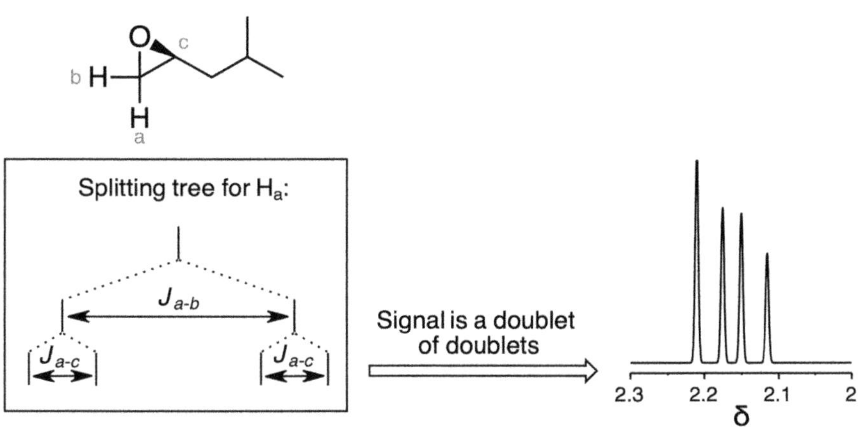

FIGURE 9.11 Splitting tree for H_a of (*S*)-2-isobutyloxirane.

8. If, due to conformational constraint, the aldehyde proton does not couple with its neighbors, then both the aldehyde proton (a) and the adjacent methylene (b) will be singlets. Aldehyde protons appear between 9 and 10 ppm. The methylene protons are adjacent to both a C=O and a C=C (of the aromatic ring). Therefore, we might expect the cumulative nature of deshielding to move them a bit further downfield than the 1.5–2.5 ppm range for protons adjacent to either one of these features individually, especially considering the electron-withdrawing chlorine on the ring.

The aryl protons (c–f) are all different but may well overlap to form a complex multiplet in the range of 6.5–8 ppm. If we were to add integration values, signal "a" would integrate for 1H, signals "c–f" for 4H, and signal "b" for 2H (Figure 9.12).

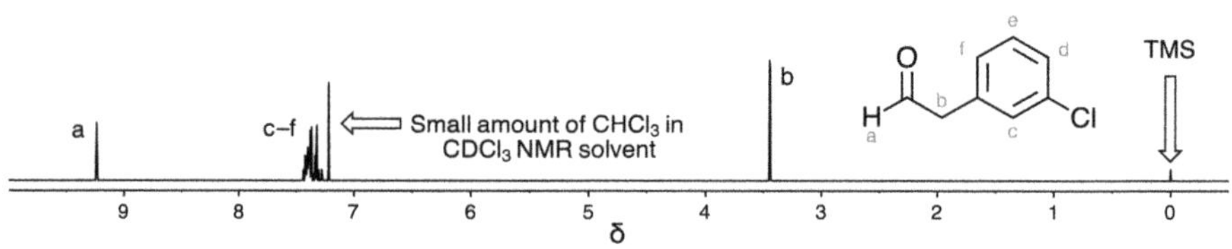

FIGURE 9.12 Proton NMR spectrum for (3-chlorophenyl)acetaldehyde.

9. In the proton NMR spectrum of crotonaldehyde (Figure 9.13), we can easily assign the aldehyde proton (~9.5 ppm) and the methyl protons (just below 2 ppm). This leaves two signals for the vinyl protons. One of these signals appears in the expected range (4.5–6.5 ppm), but the other does not. The second vinyl signal is actually in the chemical-shift range typically associated with aryl protons (6.5–8 ppm).

The resonance of crotonaldehyde helps to explain this phenomenon. There is a resonance form that delocalizes the pi electrons of the alkene, moving them toward the carbonyl. This resonance form places a formal positive charge on the β-position. Therefore, in the resonance hybrid, the β-position has a partial positive charge (δ^+). That withdrawal of electron density explains the unusual deshielding of the β-proton.

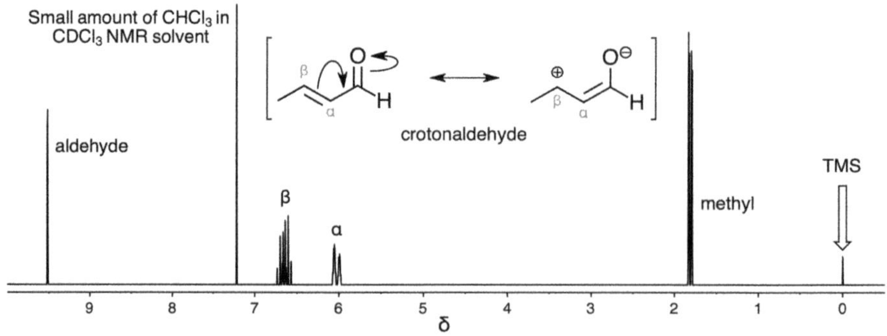

FIGURE 9.13 A labeled proton NMR spectrum of crotonaldehyde.

10. Acetophenone has resonance forms in which electron density from the aromatic ring is withdrawn into the adjacent carbonyl (Figure 9.14). This places formal positive charges on the positions *ortho* and *para* to the substituent. Therefore, in the resonance hybrid, these positions bear partial positive (δ^+) charges.

Consequently, we can gauge the *ortho* and *para* positions on the ring to be more deshielded than the *meta* position (chemical shift just below 7.5 ppm). Between *ortho* and *para*, it is the *ortho* position that is the most deshielded (chemical shift about 8 ppm) because, in addition to the resonance effect, it experiences greater inductive electron withdrawal by being closer to the substituent.

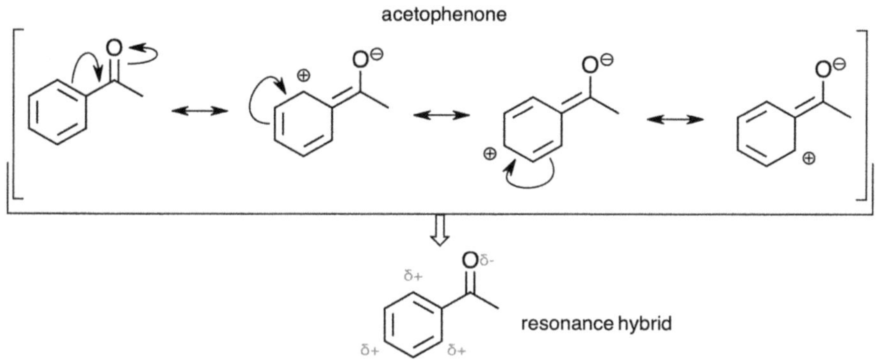

FIGURE 9.14 The resonance forms and resonance hybrid of acetophenone.

In contrast, toluene has no resonance forms. Any differentiation between its three types of aryl protons is due merely to inductive electron donation by the methyl group and is therefore less pronounced. As a result, all three types of aryl protons in toluene accidentally overlap, giving one complex and messy signal.

11. There are two salient features of the given carbon-13 spectrum. The first is the presence of a carbonyl carbon at around 210 ppm. The second is that, since there are only three signals for five carbons, the molecule must be highly symmetrical.

It would not be possible to construct a symmetrical enough five-carbon aldehyde to yield three ^{13}C NMR signals of these relative heights. Therefore, compound A must be a ketone. The most symmetrical ketone we can construct with the formula $C_5H_{10}O$ is 3-pentanone. 3-Pentanone fits the bill because, in addition to the carbonyl carbon, it produces only two other signals. The equivalent methylene groups (b), being closer to the electron-withdrawing carbonyl, are a bit more deshielded than the equivalent methyl groups (a) (Figure 9.15).

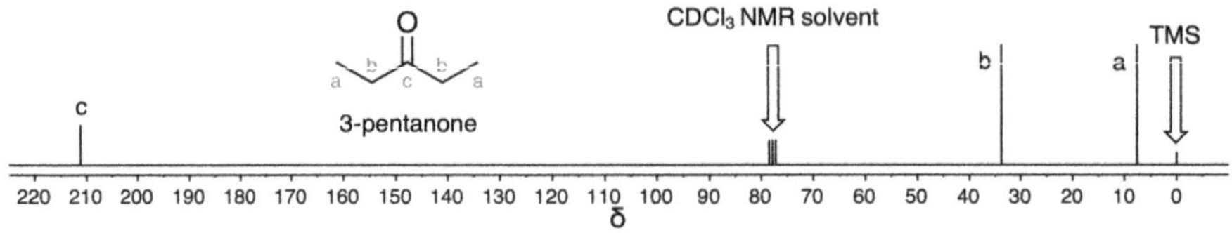

FIGURE 9.15 A labeled carbon-13 NMR spectrum of 3-pentanone.

12. The carbon-13 NMR spectrum contains enough signals between 100 and 150 ppm (Figure 9.16) that we should immediately suspect the presence of an aromatic ring. In fact, given the extent of unsaturation revealed by the molecular formula, an aromatic ring conveniently explains the ratio of carbon to hydrogen.

Additionally, the spectrum shows only one signal below 100 ppm. That rules out the possibility of the remaining two carbons being in the form of an ethyl group. The methyl and methylene of an ethyl group would have to be chemically different, and therefore, we would see two signals below 100 ppm, instead of just one.

This leaves us with an aromatic ring bearing two methyl groups. There are three constitutional isomers possible: *ortho*, *meta*, and *para*-xylene. In Figure 9.16, the different types of carbon atoms are labeled for each. *Para*-xylene would produce only three signals for its three unique types of carbon atoms. *Ortho*-xylene would produce four signals. Only *meta*-xylene produces five ^{13}C signals. Therefore, compound B must be *meta*-xylene.

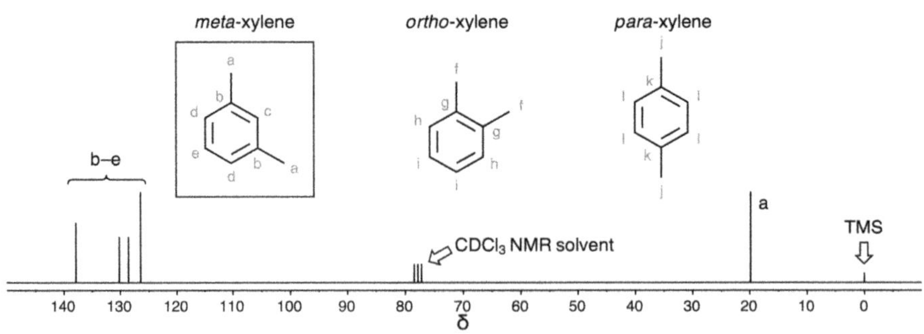

FIGURE 9.16 A labeled carbon-13 NMR spectrum of *meta*-xylene.

13. Given that compound C has the formula $C_6H_{13}Cl$ but only three proton NMR signals, it must be a somewhat symmetrical molecule. There are a fair number of constitutional isomers having this molecular formula, so drawing them all, hoping to find a match, is probably *not* the most efficient

approach. Instead, we should try to deduce structural features from the spectrum and stitch those pieces together to construct a complete molecule.

The signal integrating for three hydrogens is the simplest to decode. It denotes a methyl group ($-CH_3$) with no neighboring hydrogens.

The signals integrating for four and six hydrogens are slightly trickier to analyze. If a carbon had four hydrogen atoms, it would simply be methane (CH_4), which is a complete molecule all by itself. So, this cannot be what the signal means. Furthermore, a single carbon atom simply cannot have six hydrogen atoms because it would then exceed the octet rule.

Therefore, these signals must indicate chemically equivalent groups. The easiest way to get a signal integrating for four protons is to have two identical methylenes ($-CH_2- \times 2$), and the most straightforward way to obtain a signal for six protons is from two equivalent methyl groups ($-CH_3 \times 2$). Thinking about these fragments more holistically, they likely comprise two chemically equivalent ethyl groups ($-CH_2CH_3 \times 2$). This is supported by their splitting patterns. The methylenes have three neighbors, making them a quartet, and the methyls have two neighbors, thereby appearing as a triplet.

As you amass fragments, it is always important to check their total against the molecular formula. So far, we've identified an isolated methyl ($-CH_3$) and two equivalent ethyl ($-CH_2CH_3 \times 2$) groups. That's a grand total of five carbons and thirteen hydrogens. Comparing this to the molecular formula ($C_6H_{13}Cl$) shows one carbon and a chlorine atom, for which we must still account.

The carbon cannot have any hydrogens because there are no more proton NMR signals and no more hydrogens in the formula, so it has four available bonds. Conveniently, we can attach all four fragments (the methyl, the two ethyls, and the chlorine) to that carbon to construct a complete molecule: 3-chloro-3-methylpentane, which must be compound C (Figure 9.17).

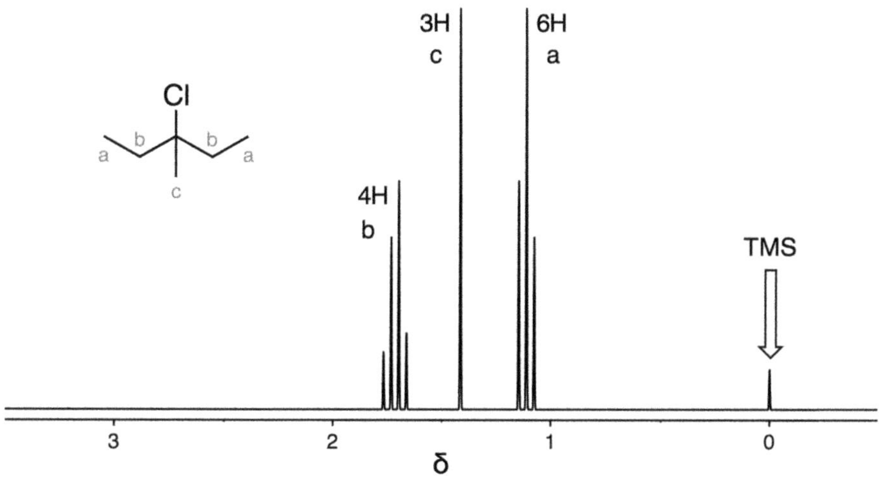

FIGURE 9.17 A labeled proton NMR spectrum of 3-chloro-3-methylpentane.

14. Before we forge ahead into the weeds, let's look at some big-picture clues in the spectra. There are six ^{13}C signals, and they all fall between 100 and just above 150 ppm. At first glance, that might seem like just an aromatic ring, but remember that the formula (C_8H_8O) contains eight carbons. If there are only six ^{13}C signals, there must be at least some symmetry. With that in mind, we then notice that two of the ^{13}C signals (at ~115 and 128 ppm) are roughly twice as tall as the others. Therefore, we could account for the signals in this region using an aromatic ring with some symmetry and an alkene. An aromatic ring (six carbons) and an alkene (two carbons) would account for all the carbon atoms in the formula, so let's move on to the proton NMR spectrum.

The proton NMR contains two aryl signals (6.5–8 ppm), each a doublet integrating for two hydrogens. This provides an important clue about the aromatic ring: it must be *para*-disubstituted.

There is one broad signal (~5.6 ppm) integrating for a single hydrogen that should be our hydrogen-bonding hydroxyl proton.

Between 4.5 and 6.5 ppm, there are three additional signals that, based on their chemical shift, indicate vinyl protons. If you place three hydrogens on an alkene, there's only one remaining valence. We must use that for a bond to the aromatic ring, and then the hydroxyl group can be placed as a phenol on the other end of the ring, yielding 4-vinyl phenol as the structure of compound D (Figure 9.18).

Though we didn't really need to pick apart the multiplicity of the vinyl protons in order to solve the problem, let's take a look at it. Protons "h" and "i" are chemically different because one (i) is *cis* to the ring, while the other (h) is *trans* to it. Geminal neighbors on a sp^2 carbon exhibit weak coupling of 0.5 to 3 Hz. On the lower end of that range, the coupling may not even be visible. That's why the signals for "h" and "i" appear only as doublets. In each case, we can see the coupling to the vicinal neighbor (f) but not to the geminal neighbor. Proton "f," however, is split into a doublet of doublets because it experiences *cis* (8–12 Hz) and *trans* (14–17 Hz) coupling to its two vicinal neighbors (h and i).

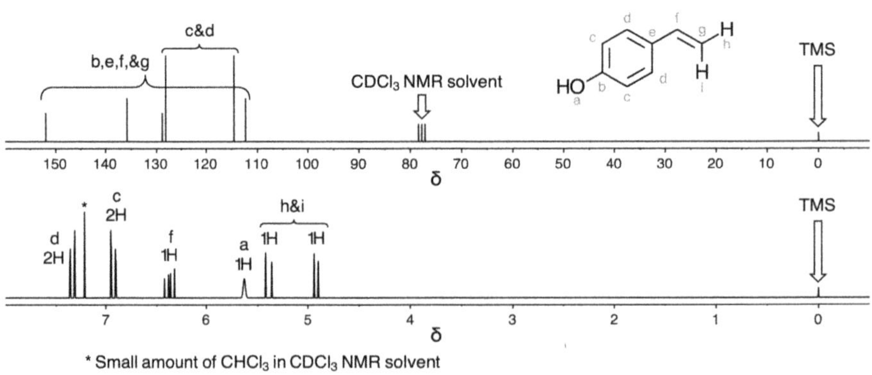

FIGURE 9.18 Labeled carbon-13 and proton NMR spectra of 4-vinylphenol.

Finally, we can return to the aromatic region and use resonance to explain why protons "c" are more shielded than "d." Delocalization of one of the hydroxyl group's lone pairs into the ring will place a negative charge at "c" (but not at "d"), causing this site to be more electron-rich and therefore more shielded.

15. You may sometimes find it helpful to look back and forth between the spectra, rather than fully analyzing one before moving on to the other. For instance, compound E's ^{13}C spectrum shows four signals between 100 and just above 150 ppm. Its proton NMR spectrum shows two signals in the aryl region (6.5–8 ppm) integrating for a total of five protons. Taken together, these clues suggest the presence of a phenyl group (i.e., a monosubstituted benzene ring). It would possess five aryl protons, and its symmetry would cause there to be only four signals (rather than six) in the 100–150 ppm range of the ^{13}C spectrum.

The phenyl group accounts for a large chunk (six carbons and five hydrogens) of our molecular formula ($C_9H_{10}O_2$). Only three carbons, five hydrogens, and both oxygen atoms remain. One of these carbons is a carbonyl carbon because it produces a signal at ~180 ppm in the ^{13}C spectrum. The other two carbons appear below 100 ppm and are fairly shielded with chemical shift values less than 30 ppm.

The proton NMR has only two remaining signals. The interpretation of NMR spectra rewards effort. With more experience, you begin to recognize patterns that facilitate the problem-solving process. For instance, we have now seen multiple times that an ethyl group yields a quartet integrating for two hydrogens and a triplet integrating for three hydrogens. The only variable is how deshielded the methylene's signal is, which depends upon what it is bonded to. In this case, the methylene signal (b, Figure 9.19) falls in the range 1.5–2.5 ppm, suggesting that the methylene is adjacent to a double bond (H–C–C=C or H–C–C=O).

If we place the ethyl group on the phenyl group, there will be no place left for the carbonyl and the last oxygen atom. However, if we place the ethyl group on the carbonyl carbon, then we may place the last oxygen between the carbonyl and the phenyl group, making an ester and giving compound E: phenyl propionate.

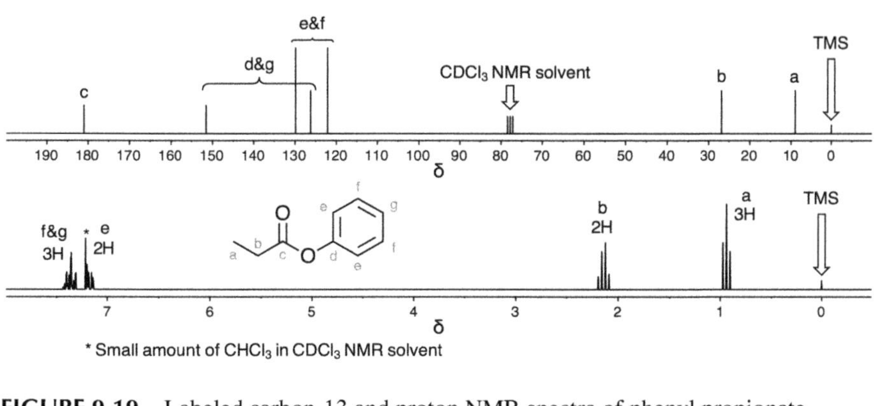

FIGURE 9.19 Labeled carbon-13 and proton NMR spectra of phenyl propionate.

16. The carbon-13 spectrum shows that compound F has one carbonyl carbon. The remaining four signals fall below 100 ppm. However, that's a total of only five signals for the six carbons in the molecular formula ($C_6H_{13}NO$). Therefore, we can deduce that two of the carbon atoms must be identical, causing only one signal between them—likely the most shielded signal (a, Figure 9.20) since it is roughly double the height of the others. For two carbons to be identical, either the molecule must be highly symmetrical (not likely given the rest of the spectral information) or the methyls must branch from the same carbon, which suggests an isopropyl group [$-CH(CH_3)_2$].

In the proton spectrum, the methyls of the isopropyl group should yield a doublet integrating for six hydrogens. Initially, it does not appear that we have such a signal, but let's consider the signal at ~1.2 ppm more closely. First, it integrates for nine protons, but a *tert*-butyl group, which is suggested by that integration, should be a singlet. The inset in Figure 9.20 shows that this signal is actually two overlapping signals because we do not have the tallest spikes in the center. This could actually be an overlapping doublet and triplet. This signal is an overlapping doublet integrating for six hydrogens (the methyls of our isopropyl group) and a triplet integrating for the residual three hydrogens.

Now that we feel more confident with our assignment of an isopropyl group because we've found its methyls (a), let's find its methine. That should be a multiplet integrating for one proton. We have such a signal at 2.5 ppm, right on the cusp of two chemical shift regions.

What remains on the right half of the proton spectrum is a quartet integrating for two hydrogens and a triplet integrating for three hydrogens. As we've seen previously, that's the calling card of an ethyl group ($-CH_2CH_3$).

We've now identified two large fragments: an isopropyl [$-CH(CH_3)_2$] and an ethyl ($-CH_2CH_3$) group, along with the carbonyl carbon (C=O). What remains in our formula is nitrogen and one more hydrogen atom. The proton signal just above

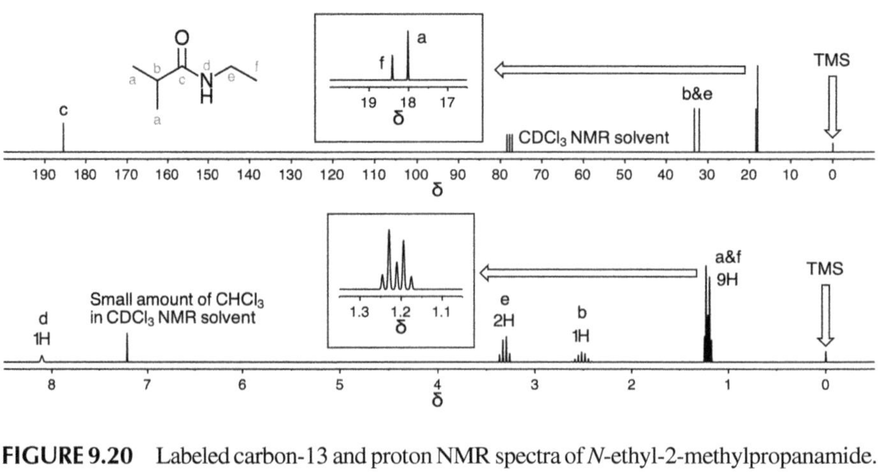

FIGURE 9.20 Labeled carbon-13 and proton NMR spectra of *N*-ethyl-2-methylpropanamide.

8 ppm is broad, suggesting hydrogen bonding. Since it also integrates for one hydrogen, it denotes an NH group.

As we seek to assemble all of our fragments, the only remaining question is whether the ethyl or isopropyl should be bonded to nitrogen. Since the ethyl's methylene protons (e) are more deshielded than the isopropyl's methine proton (b), it is the ethyl group that is bound to nitrogen. The isopropyl is attached to the carbonyl carbon. This gives compound F as *N*-ethyl-2-methylpropanamide.

17. The ^{13}C spectrum shows that compound G has no sp^2 hybridized carbons (i.e., no alkenes, aromatic rings, or carbonyls). All its signals fall below 100 ppm; however, two are noticeably more deshielded than the others. The inset shows that one of these actually falls in the midst of the three spikes of the deuterated chloroform ($CDCl_3$) signal. The significant deshielding of these two carbons suggests that they might both be bonded to the oxygen atom, meaning that the functionality would be an ether.

However, when we examine the proton spectrum, there is only one signal in the range of 2.5–4.5 ppm for protons of the type H–C–O. This indicates that only one of the two carbons bonded to oxygen has protons. The other does not. The one signal in this region integrates for two hydrogens and is a triplet, so it must be a methylene (–O–CH_2–) with two neighbors (c, Figure 9.21).

Of the remaining three proton NMR signals, one integrates for nine hydrogens. This correlates with the three equivalent methyl groups (a) of a *tert*-butyl [$(CH_3)_3C$–] fragment. The central carbon of the *tert*-butyl group (b) has no protons, so it could be the other carbon connected to the oxygen atom. At this point, we have the following fragment: $(CH_3)_3C$–O–CH_2–.

What remains in the proton NMR spectrum is a set of two signals, integrating for two and three hydrogens. That would logically be an ethyl fragment. Installing the ethyl group onto the fragment we constructed earlier completes the structure of compound G: $(CH_3)_3C$–O–$CH_2CH_2CH_3$, which is 2-methyl-2-propoxypropane.

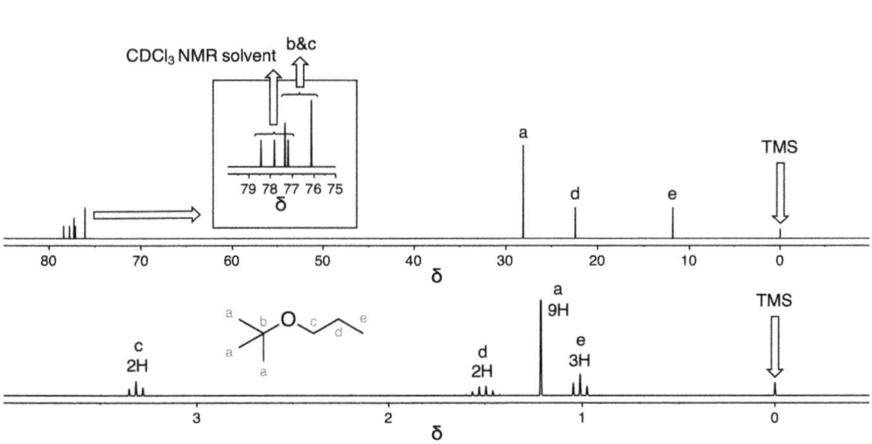

FIGURE 9.21 Labeled carbon-13 and proton NMR spectra of 2-methyl-2-propoxypropane.

Notice that the protons of methylene "d" are split into a multiplet because they have several neighbors.

18. Compound H's ^{13}C NMR spectrum shows clear evidence of a carbonyl carbon (i, Figure 9.22). One inset in the proton NMR spectrum shows a carboxylic acid proton (j). Therefore, the carbonyl is specifically that of a carboxylic acid ($-CO_2H$).

The carbon-13 spectrum also suggests an aromatic ring, due to the six signals between 100 and 150 ppm. Note that the inset in the ^{13}C spectrum more clearly shows the three slightly taller signals, two of which are quite close in chemical shift but nonetheless distinct.

The aryl region (6.5–8 ppm) of the proton spectrum shows two signals (g and h) that initially look like doublets. Upon closer inspection of the inset expanding this region, those signals are actually revealed to be doublets of doublets. One coupling constant is large and is therefore likely representative of *ortho* coupling. The second coupling constant in each case is smaller, suggesting a longer-range *meta* or *para* coupling. Taken together, these features indicate a trisubstituted aromatic ring with two adjacent (*ortho*) substituents and the third *meta* to one substituent and *para* to the other.

One of the substituents on the ring could be the carboxylic acid. The other two must be deduced from the remaining proton and carbon-13 signals. The easier of the two is a methyl group with no neighbors (k) shown by the signal just above 2 ppm that integrates for three hydrogens.

The last substituent is an isopropyl group, with its distinctive doublet integrating for six hydrogens (at 1 ppm), denoting two identical methyl groups having a

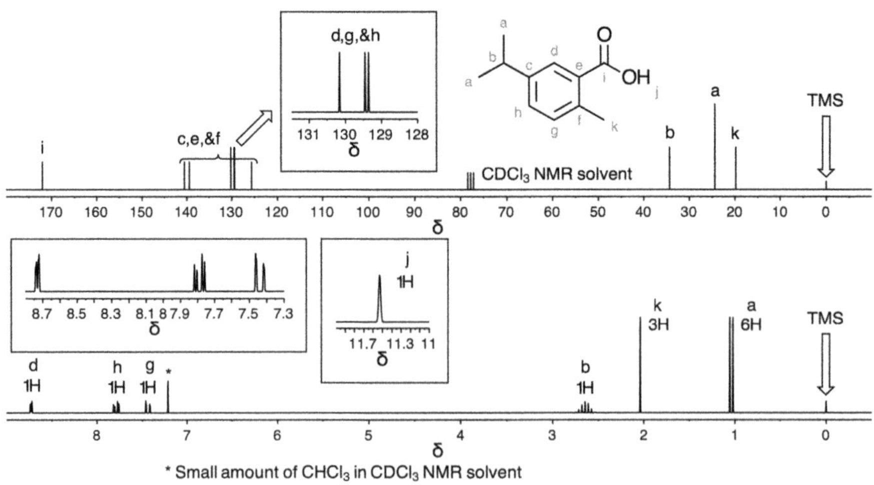

FIGURE 9.22 Labeled carbon-13 and proton NMR spectra of 5-isopropyl-2-methylbenzoic acid.

single neighbor, as well as the multiplet integrating for only one hydrogen just above 2.5 ppm.

It's worth noting that there are several constitutional isomers having many of the features we've discussed (Figure 9.23). However, some of these isomers are more consistent with the data than others. Let's examine resonance and inductive effects to find out which answers are the best. Resonance of the ring with the carboxylic acid will put partial positive (δ^+) charges on the positions *ortho* and *para* to the acid. Additionally, the inductive electron withdrawal of the carboxylic acid is more powerful when a group is closer to it.

The most isolated proton on the aromatic ring (signal d) is also the most deshielded. As such, it should be *ortho* to the carboxylic acid, where it would experience significant electron withdrawal due to both resonance and induction. This eliminates isomers 2 and 3 (Figure 9.23) from consideration.

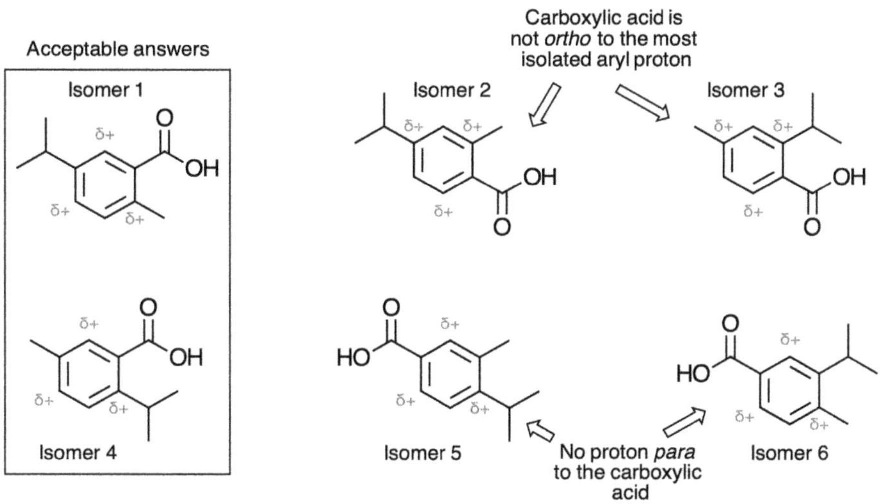

FIGURE 9.23 Constitutional isomers of 5-isopropyl-2-methylbenzoic acid.

Of the remaining aromatic proton signals, signal "h" has *ortho* and *meta* splitting because the outer sets of spikes are separated by a greater distance (larger *J* value) than in signal "g," which is more consistent with *ortho* and *para* splitting. Signal "h" is also more deshielded than "g," which tells us that this proton, which is *ortho* to one neighbor and *meta* to the other, should be para to the carboxylic acid so that it experiences electron withdrawal due to resonance. This eliminates isomers 5 and 6 from consideration since they have no proton *para* to the carboxylic acid.

This leaves isomers 1 and 4 as acceptable answers to this question. They could be differentiated using a two-dimensional NMR experiment, but that is a topic beyond the scope of this book.

19. The presence of nitrogen in a molecule as highly saturated with protons as $C_7H_{17}N$ can only be explained by an amine. However, we see no proton NMR signals that are broadened by hydrogen bonding. Therefore, compound I must be a tertiary amine (R_3N).

The three most deshielded signals in the ^{13}C spectrum must be the carbons directly bonded to nitrogen (d–f, Figure 9.24). Similarly, the three most deshielded sets of protons should be those on the carbons bonded to nitrogen (d–f). These appear not as three separate signals but as only two signals, where one consists of overlapping types of hydrogens. The singlet integrating for three hydrogens at 2.4 ppm clearly denotes a methyl group bonded to nitrogen ($R_2N–CH_3$).

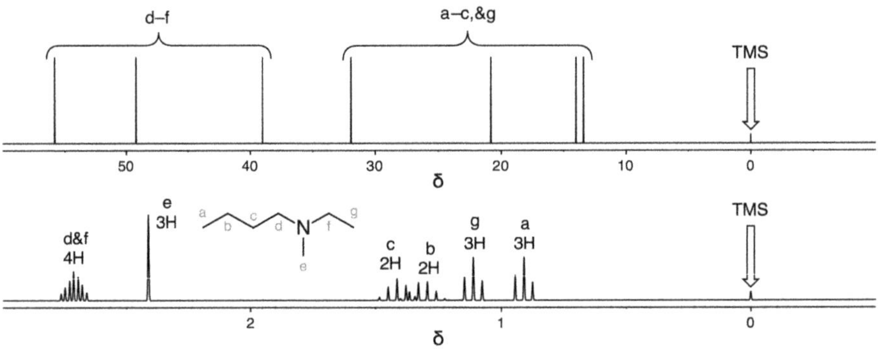

FIGURE 9.24 Labeled carbon-13 and proton NMR spectra of *N*-ethyl-*N*-methylbutylamine.

The multiplet at 2.7 ppm integrating for four indicates two similar methylene (CH_2) groups, both of which should be bonded to nitrogen, given their appearance in the range 2.5–4.5 ppm (H–C–N). This gets us to the fragment shown in Figure 9.25.

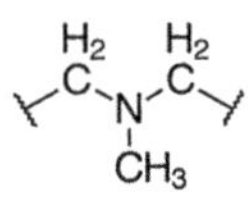

FIGURE 9.25 Fragment constructed from the three most deshielded types of protons.

The remaining proton NMR signals from 0.5 to 1.5 ppm suggest two chemically distinct methylenes (CH_2) and two chemically distinct methyls (CH_3). There would be two ways to assemble these fragments, only one of which is actually consistent with the NMR data (Figure 9.26). Compound I is *N*-ethyl-*N*-methylbutylamine.

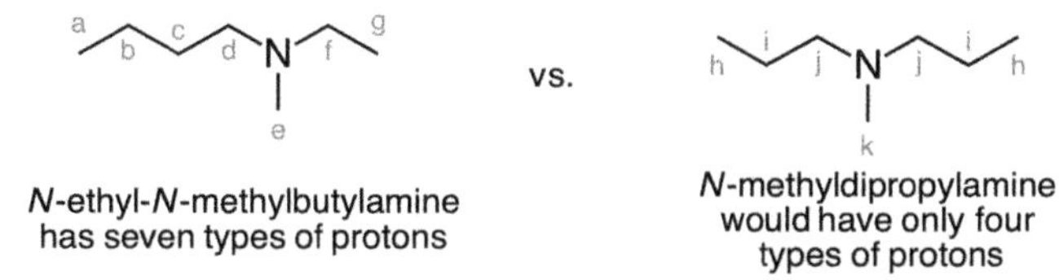

FIGURE 9.26 Two constitutional isomers that can be assembled from the final fragments.

20. Compound J has one carbonyl carbon, which appears just above 200 ppm in the ^{13}C spectrum (i, Figure 9.27). In the proton spectrum, there's a distinctive aldehyde proton just above 9 ppm (j), which reveals the specific identity of the carbonyl.

There are four signals between 100 and 150 ppm in the ^{13}C spectrum (b–e) and two signals, each integrating for two hydrogens, in the aryl region of the proton spectrum (c and d, 6.5–8 ppm). Taken together, these facts reveal the presence of a *para*-disubstituted aromatic ring. That's the only way we can obtain two doublets integrating for two protons in the aryl region of the ^{1}H NMR and four aromatic signals in the ^{13}C spectrum, where two are roughly twice as tall as the other two.

Let's pick the low-hanging fruit first as we examine the remainder of the proton NMR. There are clearly two methyl groups, since we have two signals integrating for three hydrogens apiece. One of these is a singlet (a), so it has no neighbors. The other is a doublet (g), so it has one neighbor.

If we now take stock of the fragments we've amassed and compare them to the molecular formula, we'll see that we've accounted for one carbon in the carbonyl, six in the aromatic ring, and two between the two methyls. That's a total of nine carbons. We've also accounted for the aldehyde proton, four protons on the aromatic ring, and six between the two methyls for a grand total of eleven protons. The aldehyde accounts for the sole oxygen atom. From the molecular formula ($C_{11}H_{14}O$), we can subtract $C_9H_{11}O$, leaving us with only two carbons and three protons for which to account.

This shows us something very important. There are three remaining proton NMR signals, each integrating for one hydrogen, but these *cannot* be three methine (CH) groups because we only have two carbons left. These three signals must result from diastereotopic protons of a methylene (CH_2) and a lone methine (CH).

Let's start to assemble all of our fragments. The methyl with no neighbors (a) can be attached directly to the ring. The other methyl (g) needs one neighbor, so it can be connected to the methine (f). The methine can then connect the ring and the diastereotopic methylene (h). Finally, the methylene (h) is bonded to the aldehyde to complete the structure. Compound J is 3-(*para*-tolyl)butyraldehyde.

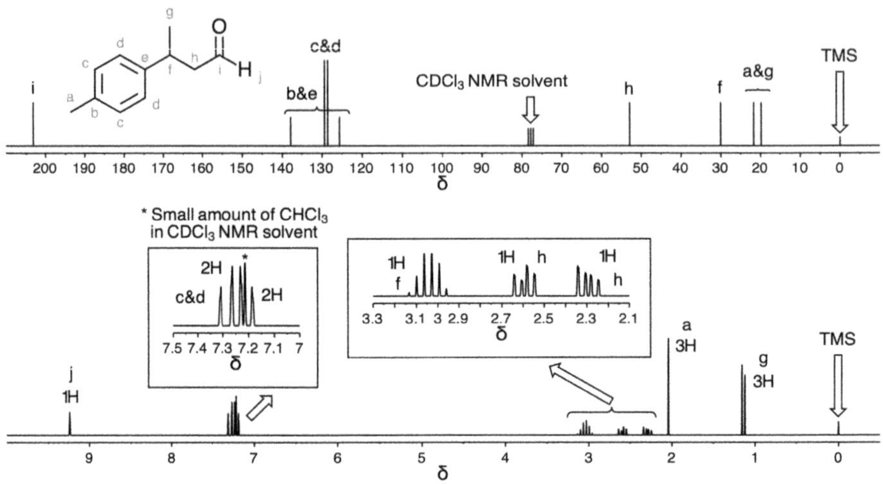

FIGURE 9.27 Labeled carbon-13 and proton NMR spectra of 3-(*para*-tolyl) butyraldehyde.

In retrospect, we can now explain some spectral features more clearly. The methine (f) is a chiral center. As such, it places the methylene protons (h) in distinct chemical environments, causing them to be diastereotopic and to yield their own individual signals (both labeled “h”).

These diastereotopic methylene protons (h) will each appear as doublets of doublets because they split each other as geminal neighbors and are also split by their vicinal neighbor (f). The methine proton appears as a multiplet because it has several neighbors (three in the methyl group “g” and two of differing types in the methylene “h”).

Although the aldehyde proton (j) technically has two vicinal neighbors, we saw previously in Chapter 8 problem 8 that aldehyde protons are often not split by their neighbors due to conformational constraints. If any such splitting is perceived, it is typically minuscule due to a very low *J* value.

Index

Pages in *italics* refer to figures

I

J

K

L

M

N

O

P

Q

R

S

T

www.ingramcontent.com/pod-product-compliance
Ingram Content Group UK Ltd.
Pitfield, Milton Keynes, MK11 3LW, UK
UKHW022121190926
475702UK00013B/1572
* 9 7 8 1 0 4 1 3 2 5 2 5 3 *